42 Rules of Produc Marketing

Learn the Rules of Product Marketing from Leading Experts from around the World

By Phil Burton, Gary Parker, and Brian Lawley

E-mail: info@superstarpress.com
20660 Stevens Creek Blvd., Suite 210
Cupertino, CA 95014

Published by Super Star Press™, a Happy About® imprint
20660 Stevens Creek Blvd., Suite 210, Cupertino, CA 95014
http://42rules.com

First Printing: March 2012
Paperback ISBN: 978-1-60773-080-4 (1-60773-080-4)
eBook ISBN: 978-1-60773-081-1 (1-60773-081-2)
Place of Publication: Silicon Valley, California, USA
Library of Congress Number: 2012932095

Trademarks

Warning and Disclaimer

Praise For This Book!

"*42 Rules of Product Marketing* provides unique insights on how to improve the effectiveness of product marketing investments and does an excellent job of collecting and organizing the best practices of thought leaders and experienced practitioners from a spectrum of product marketing functions.

Regardless of your role, organization, or background, you will gain new insights and ideas on how to improve the effectiveness of your product marketing activities."
Ken Horner, Vice President of Business Development, Calgary Scientific Inc.

"*42 Rules of Product Marketing* is an excellent compilation of easy-to-read marketing tips that cover state of the art marketing practices and provide fresh perspectives and ideas to help any business move forward."
Mike Freier, President, Silicon Valley Product Management Association

"I want to thank the authors of *42 Rules of Product Marketing* for their contribution to the product marketing profession. They have created a powerful but concise book that pulls together the insights, experience, and wisdom of over 40 marketing professionals with excellent advice for both consumer and business products. This book is an excellent companion to *42 Rules of Product Management*."
Therese Padilla, President, Association of International Product Marketing & Management

Dedication

This book is dedicated to the product marketing managers around the world, in every industry, who are committed to the success of their organizations by creating messages, customer collateral, sales tools, and everything else necessary to drive revenue for their companies' products and services.

Acknowledgments

We would like to thank the outstanding team of product marketing experts worldwide who contributed their time and wisdom to this project. Without them, this book would not have been possible. By sharing their knowledge and insights, they have helped further the profession of product marketing.

We also want to express appreciation to our families for their support and patience during the creation of this book. We used to wonder why authors always dedicate books to their families—now we know!

Contents

Contents

Over our careers, we have watched as product marketing management (PMM) has evolved into a role that is absolutely vital to a company's success. Years ago, PMM functions, if done at all, were typically included in the product management (PM) role. The problem was the constant struggle between the need to develop market requirements for the engineering team against the equally important need for product launch, message creation, sales support etc. Often, short term deadlines for revenue demands from sales would take priority to the detriment of the company's longer range issues of product and market planning. By sheer necessity, companies around the world began to separate PMM from PM functions to achieve the right balance between the need to develop the product and the need to ensure its success in the marketplace after it was released. Establishing a PMM role meant that the sales teams could be assured of having a consistent, unified, and focused approach to messaging.

The PMM role has always been important, but it became absolutely critical with the advent of the Internet. As you'll read in various Rules in this book, today's marketplaces are different because almost everything that buyers want to know about a company or product can be learned online. In the past, the sales team and the company's advertising would educate buyers in the early stages of the buying cycle. Essentially, the PR, advertising, and sales teams could manage the information available to buyers and users.

However, in today's Internet-based marketplace, with the rising importance of social media, savvy buyers expect to find the initial information for themselves from independent websites and from other customers, and don't want vendor contact until they are ready to buy. This means that the vital function of early prospect education has shifted dramatically and a PMM needs to utilize new marketing automation methodologies and tools to nurture leads until they are ready for the sales team to take them over.

Of course, despite all these changes, some basics never change. Product messaging is always about helping a prospect solve their business problem and address their needs, not simply describing product features. What has changed is how web-based messaging supplements—or in some cases replaces—the traditional product collateral so that the product marketing manager has to become a partner with the company's prospects in building a needed solution.

We are very fortunate that marketing experts from around the world took the time to share their insights on these very important topics. If you're like us, when you read their rules, you'll find reminders of fundamentals that we all need to remember as well as fresh insights and approaches to vital aspects of your job.

We sincerely hope that the rules in this book will help make your product marketing career more enjoyable and more successful! Let us know your thoughts at http://280group.com/blog/.

Phil Burton, Gary Parker, and Brian Lawley

1

Rules Are Meant to Be Broken

By Gary Parker, Product
Marketing Manager, FalconStor Software

Never forget that the underlying driver is a passion to produce product messages that resound with our well-understood buyer personas.

This rule may seem unnecessary; after all, isn't the point of having rules to share and learn from other people's experiences and research?

We believe this is a Zen type of statement that is telling us to balance in-depth marketing knowledge with the need to retain passion and creativity.

The late Steve Jobs was extraordinarily successful at breaking the rules and was famously quoted in *Fortune*[i] for saying "We do no market research.... We just want to make great products." Taking his statement literally and ignoring the good advice from Rule 7 on market research would be a career threatening move for most of us and certainly has contributed to the high rate of start-up company failures.

So, how do we reconcile this apparent contradiction? The value of Jobs' contrarian approach for us ordinary mortals has several aspects. The first is the importance of having a deep passion for our products, with our marketing tools being a means to express that passion. Jobs' own words prophetically said it well: "We don't get a chance to do that many things, and every one should be really excellent. Because this is our life. Life is brief, and then you die, you know? So this is what we've chosen to do with our life.... And we've all chosen to do this with our lives. So it better be damn good. It better be worth it. And we think it is." His insight reminds us that all the valuable marketing ideas from this book's contributors and other sources are very important, but they are tools to be utilized to accomplish our marketing goals and express our profound underlying desire to succeed.

Another aspect is the importance of deeply understanding our buyer personas to the point that they become living entities that drive our work. Again, Jobs clearly understood his buyer's persona and took full responsibility for meeting their needs: "Our DNA is as a consumer company—for that individual customer who's voting thumbs up or thumbs down. That's who we think about..."

There was absolutely nothing subtle about this statement. He told us that, as marketers, we own our personas and are fully accountable for our ability to translate their real needs into succinct, compelling marketing messages.

There are many other aspects of Jobs' apparent rule breaking legacy that can help us, but one final one is the importance of interpreting market data based on the real, underlying customer needs and not necessarily on the customer's ostensible requirements statements. At the risk of being trite, we'll close with Steve's comments on asking what people want: "So you can't go out and ask people, you know, what is the next big [thing.] There's a great quote by Henry Ford, right? He said, 'If I'd have asked my customers what they wanted, they would have told me, 'A faster horse.'" That traditionally applies to product managers, but also applies to product marketing when we use customer interaction and understanding and research to develop winning messages.

So, please enjoy and absorb the hard earned lessons that these rules represent, but never forget that the underlying driver is a passion to produce product messages that resound with our well understood buyer personas, as a key driver to marketplace success.

2

THINK BIG!, start small, *move fast...*

By Christine Heckart,
Chief Marketing Officer, NetApp

To think big is to dream big. Imagine you are at the end of the process and define what success looks like.

Everyone knows strategy drives execution. But what drives strategy? And, what kinds of strategy drive the right kinds of action, at the right times?

We are always very, very busy. When you consider all the things we can spend time doing at any moment, it is hard to discern which are the most effective in achieving specific short-term and long-term objectives. When you can't do everything, how do you choose?

The best methodology I've found for managing strategy and execution is to THINK BIG!, start small, *move fast....*

THINK BIG!

To think big is to dream big. Imagine you are at the end of the process and define what success looks like. What are you trying to achieve? What impact do you want to make on your company, your industry, or the world?

For most big endeavors this means looking three to five years into the future, though some things might need a shorter time span. The important point is to make the time horizon long enough to let you dream expansively, imaginatively, and optimistically.

Too often, we hold ourselves back from really thinking big. We get caught up in minutiae, we compromise before we even begin, we focus on the reasons our big ideas won't work. At the THINK BIG! stage, put all these hurdles aside, and give your imagination free reign.

Envision what might be true and different in the future. How would we like people to think about us, our products or services, our company? What different behaviors would arise? What new points of view?

When you're thinking big, define success in all the ways that matter, both quantitatively and qualitatively. Consider hard numbers like revenue forecasts, but also include the qualitative aspects that ultimately matter when you think deeply about "success." Describe the vision in enough detail and with enough juice that everyone involved can understand the goal and why achieving it matters.

start small

Once you have clearly defined what success looks like, identify the few things you can start on *now* to move toward that goal. Don't try to boil the ocean.

Pick a way to test the big idea and prove its value. Choose one aspect required for success and get it done. Start with something that provides early traction and credibility, gives you quick feedback, and allows you to build awareness and momentum.

Of all the possible things to do, focus first on the handful that are most linked to achieving your big idea. The leftover things might be useful later, or maybe not. Avoid diluting your efforts by chasing too much too soon. Start small, with achievable actions, and build from there. And just as important, put aside activities that *don't* link to the big idea—however interesting, fun, or compelling they may sound.

move fast....

Prioritizing and starting small doesn't mean lollygagging, however. Big ideas don't reach fruition without continuous wins, continuous progress. And they don't happen if you lose traction or move away from your main goal. If you think big, start small, but then start over, you've ultimately achieved nothing of importance.

It's human nature to get lost in the weeds, to say yes to every meeting, to devolve into a focus on management-level detail, to act for action's sake, to fall into a pattern of believing that all deadlines are equally important.

To move fast, always keep sight of your big goal, and evaluate your small actions by how well they move you there. Remind people, over and over, of the goal you are aiming at. Start every meeting with the big idea, and for every activity proposed, ask if it is the most essential activity to move toward that idea. Then, reward and reinforce each small step that reaches closer to your goal.

THINK BIG!, start small, and *move fast....* Use this methodology to help you define success—and mobilize and motivate your team to achieve it.

3 Learn from Your Customers' Digital Body Language

By Steven Woods,
CTO and Co-Founder, Eloqua

Each of these actions, the user's "digital body language," can give you insight into who is just kicking the tires, and who is likely to take a step forward and upgrade.

Today's buyers are different than buyers a decade ago. Everything they need to learn about your company, your product, and your team, they can learn online. They are not out there learning by having your sales team educating them, they are learning by sharing perspectives with their peers, or getting hands on experiences with your product itself.

This change leads to an interesting challenge and a great opportunity for today's product marketers. The product itself is a key part of the education cycle. From free trials, to entry level products, to active usage of your main product, every experience a user has with your product is an opportunity to help you understand where they are in considering your offering, and you can guide your messaging strategy accordingly.

The challenge for product marketers is to rethink a communication strategy around the buyer's online experience. Buyers should no longer be segmented by demographics or firmographics, or communicated with in large outbound campaigns. Buyers must be understood in terms of the activity indicators and psychographics that tell you where that person is in their own unique consideration cycle.

The data we now have on buyers is as broad as it is insightful. Have they logged in for the first time? Have they configured their account? Have they taken the first action that you want them to in the product? Have they tried to use a feature that was disabled because it was only available in a higher level version of the product? Each of these actions, the user's "digital body language," can give you insight into who is just kicking the tires, and who is likely to take a step forward and upgrade.

Each communication path you have with potential buyers should leverage this insight by helping that individual buyer try more, learn more, or consider purchasing your product. Without this insight, using only the broad market segments that define industries and roles, you are left with undifferentiated mass communication and are merely hoping to connect the right message with a buyer at the right point of consideration.

Taken one level further, the insights that the sales and marketing team gain on each person and organization who is considering your product could easily be fed back into the product experience itself and used to focus messaging, offers, and education within the product. It is also vital to the sales team, who can establish a strong working relationship by referencing the prospect's activities such as white papers or free demos instead of having to make a blind inquiry and trying to extract information that the prospect has already provided to your company.

For today's buyer, the product itself is the core of the buying process. As a product marketer, understanding this point and building a messaging plan that leverages an understanding of where a buyer is in their consideration cycle, based on their actions, is core to product marketing that successfully guides potential buyers from the earliest stages of awareness and education through to consideration, purchase, and continued growth as a client.

4 Help Your Prospect Know "What's In It For Me?"

By Russell Gould, Director, Product Marketing, Guidance Software

To be relevant, content needs to follow the WIIFM (whiff 'em) or "What's In It For Me?" rule; don't pound prospects with your messaging.

As a product marketer you're responsible for your product's success. This means consistent revenue growth. While there are many success factors to growing revenue, demand generation is certainly a key component. Demand generation is not just about programs, events, sending emails, etc. It is about guiding potential customers through their journey toward purchase. Holistically, it leverages the marketing toolkit, not just one facet. It is essential to provide prospective customers the information they need throughout their buying cycle, not just the sales cycle.

Product marketers understand the sales cycle and associated funnel. What most don't realize is that there is also a marketing funnel. Working the marketing funnel is crucial to effective demand generation, particularly for complex selling cycles where 50 percent of the leads in year two come from prospects generated by the previous year's marketing activities, even when the sales cycle is only six months long. This means that marketing needs to interact or "have a conversation" with prospective buyers for twelve to twenty-four months, even though the sales cycle is much shorter.

Successful demand generation is one-to-one marketing on a mass scale. To accomplish this, you need software. There is now affordable software that automates the processes and everything marketing does to interact with prospective customers. Software provides the necessary qualitative and quantitative metrics the demand generation team needs to analyze and adjust, and be much more scientific about the process.

Collaboration is essential. Don't just ask for a webinar or another email to be sent; these are just random acts of marketing. Create a plan with the demand generation team that maps to the marketing funnel and addresses prospective customer needs across the buying cycle. In larger organizations, take a leveraged approach and collaborate with PMM peers to create top of funnel programs for multiple product lines, not just one.

To be relevant, content needs to follow the WIIFM (whiff 'em) or "What's In It For Me?" rule; don't pound prospects with your messaging. It's about them, not your product. This is basic marketing; however, time and time again marketing professionals forget this. Also, "It's about them" doesn't mean target segments; it means personalization to the individual, as noted in Rule 3 above. Developing buyer personas is not enough; the demand generation team needs to understand what is relevant to particular individuals. Classic assets such as white papers, buyer's guides, etc., are important, however, the team also needs smaller, more consumable assets on very specific topics. We live in a Twitter world; a three page paper on a specific topic instead of 10 pages on three topics will help your demand generation team learn what is relevant to that individual.

Less is more. Sending too much or asking too many questions increases the risk that prospects disconnect. Demand generation results increase with more specific individual information. What time of day do prospects want information, how frequently, via what medium, bulleted or paragraph format? Until your marketing team better understands the individual, communicate less. They need to deliver the right message, via the right medium, at the right time to the right person.

It must be easy for prospective customers to get the information they need. Is all content behind gated forms? What paths do prospects go through to get the information they need? Is content customized to the individual? How considerate are you of their time? Do they have to fill out a form every time they come to your website for content? Do they get asked questions you already have the answer to?

Effective demand generation is a key component to your product's success. While you don't have to be a demand generation expert, you can provide leadership to your marketing team to have the right mindset and focus on the individuals that buy your product. Keep driving this point home to your marketing team—your competitors are. Your product's success depends on it!

5 Follow the New Rules of PR!

By Don Jennings, Vice President,
Client Services, Lois Paul & Partners
Public Relations

The previous focus on stand-alone product features is less relevant today with the product marketer now talking more about the bigger picture.

One of the biggest press relations problems for product marketing managers is that the media landscape has changed dramatically over the last five to six years. There are now fewer publications and even fewer reporters, so there is less opportunity to engage media about product specific pitches.

However, that comes with good news in several areas. With this shift, there now are more freelance reporters that you can work with and, more importantly, there has been an explosion of social media channels that offer new ways to engage more directly with your target audience in a one-on-one approach using tools like blogs, Twitter, user communities, LinkedIn, Google+, etc.

Traditional media relations consisted of spending perhaps 80 to 90 percent of the media time on perhaps eight to twelve critical reporters and editors, but now there is more of a 50–50 or 60–40 balance in the social media/traditional media engagement. This has meant a change in media strategy. The previous focus on stand-alone product features is less relevant today with the product marketer now talking more about the bigger picture: what is the impact for the user and how the product helps productivity, enhances task automation, and improves efficiency. Also, what are industry trends and areas of focus that this product ties into so you can tell a broader story of your product as an accelerant for industry trends?

Of course, there are new rules for a new environment. In social media channels, authenticity and transparency are vital to maintaining credibility. People want a clear understanding of your affiliation so it is an authentic experience. It's ok to say you're from the vendor and here is your input. You and your company take that approach because it's right and it's the best way to reach the influencers with a wide following; but, by the way, in the United States the FTC has some strict guidelines on bloggers disclosing any affiliation or payment for what they say.

When working with a user community, always start by monitoring your audience first so you can understand what the audience is looking at and what their hot buttons are so that you can engage in an impactful way. That allows you to become a key member and influencer in those communities.

You have major resources within your own organizations such as sales engineers, product development, product management, and others, but you need an open approach that recognizes that it's not a "Wild, Wild West" out there. Everyone who participates needs to realize that they are ambassadors of the brand and the corporate philosophy while still maintaining the authenticity, legitimacy, and transparency that is a key part of social media interactions. As manager of the brand, you'll need to set guidelines for your internal contributors and help them understand that they still have obligations to your employer. This will include confidentiality for certain types of information and human resource policies as well as brand appropriate responses. You should work closely with your social media manager, the PR team and other people who monitor and oversee those types of programs. A good approach is to start with internal blogs where you can nurture the type of dialog you are looking for.

Of course, the whole point of this effort is to monetize your product or service at some point, so you need to be able to measure the effectiveness of these programs from an engagement and influence perspective as well as driving eyeballs to the website and downloads. You want to report back that your involvement in social media activities is ultimately driving higher ROI. For that, see the other rules on web and social media.

6 Work Effectively with Analysts

By John Armstrong, Silicon Valley-Based Consultant, Previously Gartner Vice President and Chief Networking Analyst

The ability to work and communicate effectively with industry analysts can make a significant difference in the analyst's perception of your product or service.

If you are recognized by your company as a subject matter expert in your product marketing role, you can expect to eventually participate in meetings with industry analysts, usually as part of the typical external launch process. Industry analysts function as experts within a product or market segment, have a unique perspective on the competitive landscape and market trends, and can be key influencers when your product or service is being considered by a prospect. Reports written by respected industry analysts are widely read by existing and potential customers, investors, resellers, as well as other vendors. Analysts are also tapped by industry press who are looking for an expert opinion or quote for an article.

Having spent nearly four years as a Vice President and Chief Networking Analyst at Gartner, I can assure you that the ability to work and communicate effectively with industry analysts can make a significant difference in the analyst's perception of your product or service, and ultimately, how your company is viewed in the broader competitive environment. Successfully interacting with analysts is a critical skill for every product marketing professional to master.

Unless you work for the tiniest of start-ups, the initial contact, analyst relations management, and meeting arrangements will be handled by the PR/AR person representing your company. They will have already identified the analyst(s) associated with your product and/or market. Your job will be to talk expertly about your product and explain what makes it different and better than existing solutions in the market, all in an attempt to positively influence the analyst.

There are right ways and wrong ways to do this. Here are a few helpful tips:

- **Take a look at what the analyst has written lately.** This will help you understand his/her perspective on the market and competition, and enable you to better focus your discussion.
- **Keep your presentation deck brief, polished, and to the point.** You have limited time to present your story. Highlight what is unique about your product or service, and what distinguishes it in the market. Stick to the facts, avoid hyperbole, and don't give a sales pitch.
- **Be completely honest when responding to a question from the analyst.** The analyst may already know the answer and is simply looking for confirmation. Don't try to obfuscate or hide unflattering news, as the analyst will find out anyway.
- **Know your stuff.** Analysts will sometimes ask detailed technical questions, and it is best if you can respond in real-time during the call.
- **Listen.** If the analyst suggests that your product or service needs improvement in specific areas, pay attention. Whether you agree or not, it is virtually certain that many of your prospects will.
- **Ask questions.** Without exceeding the bounds of NDAs, analysts like to share their knowledge and insight...That's what their clients pay them for. Maybe you will learn something new about your competition!

Some other things to keep in mind: You don't have to be a client to meet with an analyst firm. Any vendor can request a briefing with an analyst, but don't automatically expect that the analyst will write about your product unless it is a game-changer, or it falls within a current focus of research for the analyst.

Analysts are particularly interested in meeting with companies that can move markets, either with innovative or disruptive new products and services, or through the weight of established solutions. There are no guarantees, but by observing these guidelines, you can improve your odds for product (and career) success and be on your way towards developing productive relationships within the analyst community.

7 Business Research Is about Decisions, Not Questions

By Al Scharer, President & Principal Consultant, Filigree Consulting

Define the decisions that you need to make, then define the research tools you will use to make the decisions.

From a product marketing perspective, research is about decisions. Research is a systematic approach to gathering data that supports your decision making. There are several steps to create an efficient research methodology that will help you create on-target messaging.

First: Define the decisions that you need to make and then define the research tools you will use to make the decisions. Next, determine the way you are going to gain consensus on the information you develop, the information you need to populate the tools, and finally the data you will need to derive information. This is the path to defining actionable research.

Diagrammatically, research projects are defined as follows:

Decisions ⟶ Tools ⟶ Agreement ⟶ Information ⟶ Data

The research project/decision making is executed in the reverse order:

Data ⟶ Information ⟶ Agreement ⟶ Tools ⟶ Decisions

Let's use a simple example to explain. Suppose you need to make a decision about how to spend your marketing communications budget across a set of segments.

PMM Decision: I have a fixed sum of money and five market segments, how should I allocate the money to maximize revenue? For example, what campaigns should be emphasized? Which segment gets the most focus and funding in the launch?

Tools: A portfolio model linked to resource allocation. The information you need to populate the model includes: market attractiveness (e.g. opportunity, growth rate, competitive intensity, etc.) and organizational strength (solution fit to segment needs, effectiveness of channels, current level of success, etc.) for each of the segments.

Agreement: This is where the chain most frequently breaks down. In over twenty years of business research, we have seen many more arguments about what the facts are as opposed to useful dialog about what to do about the facts. If you can't agree about the facts (e.g. that a new competitor is a real threat), you are unlikely to formulate and execute a strategy that protects your business from the threat. Get your team together and agree on the facts that are necessary to decisions, then decide as a team.

This step in the process is also important as it provides a forum for integrating other research and knowledge held by the team. We often remind our clients that the research we are working on is simply another set of information that needs to be shaped by their experiences and wisdom.

Information: Simply put, information is data in context. It's a matter of distilling, refining, enhancing, and cross-referencing raw data to make it clearer and easier to internalize. Using our budget example, we might compare the growth rate of segments in the market to adjacent segments. The fact that a segment is growing at 12 percent is interesting data. That a segment is growing at 12 percent, while the total market is growing at 5 percent, is valuable information.

Data: We collect data about segment growth, size, etc. This is generally done in the traditional research path of secondary, followed by qualitative (focus groups, interviews, etc.) and then quantitative surveys.

The execution path is simple; you reverse the definition path. Get the data, refine it into information, meet to agree on the facts, populate the models, and make decisions. Voilà: actionable research!

8 Have Courage at Launch

By Brian Lawley,
CEO & Founder, 280 Group LLC

Oftentimes the courage they had to build the product turns into denial about what it really takes to build a viable business.

Sure you have a great product idea and the guts to build it, but do you **really** have a realistic picture of what it is going to take to bring it to market? Do you have the courage to spend enough to launch your product and create significant momentum?

My company, the 280 Group, works with many companies that are trying to launch products. Oftentimes clients first contact us when they are three to six months away from launching a product to help them plan how they are going to reach customers and make sales. Or they'll bring us in shortly after the product has been released and they can't figure out why sales have not begun to take off.

These companies will have spent tens or hundreds of thousands (sometimes even millions) of dollars, and months or years developing their product. They have a huge sunk cost that has been incurred to get to market. They'll ask us to create a marketing plan to reach their target market as cost effectively as possible so that they can generate qualified leads and make sales. Their attitude is usually that, "We'll be so smart about marketing (like we were with development) that we will get things going without having to spend much, and then we can fund future efforts from the initial sales." But when the rubber hits the road and the company has to invest enough time, resources, and money to get sales and revenues going, oftentimes the courage they had to build the product turns into denial about what it really takes to build a viable business.

One company that we worked with had spent $3 million doing a management buyout and bringing a new product to market, with the hope of turning the company around and generating $8 to $10 million in sales. When we proposed a ridiculously low marketing budget of $50 thousand (with some very creative leveraged marketing ideas). It took them eight weeks of internal debate to finally decide that they weren't quite sure enough about the product and market to spend the money. Guess what? After all that work they didn't come close to making their revenue projections.

What's the lesson here? Start your marketing and go-to-market planning very early—even before you make the commitment to build the product. Get real about what it will cost you, and get a seasoned marketing person involved to help you create a ballpark estimate. Take that estimate and double it as your working figure.

Sure, you might be able to come up with the next completely viral product that requires no marketing, but these are few and far between. When you are making the decision about whether to develop a product ask yourself, "When we have built the product, used up more funding than we dreamed we would to do so, and are ready to really try to get the revenue engine going, will we have the Marketing Courage to make it happen?"

If the honest answer is no, don't do it.

The moral of the story? Go big or go home.

9 Make Your CFO a Social Media Fan

By Paul Dunay, Chief Marketing
Officer, Networked Insights

Social media definitely pays off, but you've got to focus on the approach, not the tools.

Social Media has hit the mainstream, but the concern on every product marketer's mind has been how to justify the ROI, especially at budget time. The financial decision-makers might harbor unspoken thoughts like, "We spend big bucks on those developers, customer support, and sales engineering people. Why are you wasting the company's money giving them geek toys when they should be doing real work for us?"

If you're concerned about social media cost justification, you're not alone. A recent Chief Marketing Officer (CMO) survey[ii] found that only 20 percent of CMO's felt social media produced measurable ROI and 62 percent hoped it may someday. If CMO's lacked ROI confidence, then just imagine what a CFO or CEO survey might have told us!

Social media definitely pays off, but you've got to focus on the approach, not the tools. The classic product marketing approach to social media is treating it like just another broadcast media channel by posting press releases, events, videos, etc. and expecting instant results. That's a good start, but the results you need to justify your efforts will come by taking a holistic company approach.

Here are some real-life examples. I am sure you can find others when you start thinking this way.

Improved Customer Support

Avaya found that social customer support is a powerful way to reduce customer churn and increase retention rates. Listening for a customer issue, responding, and solving their problem in minutes using social media provides an "exceptional customer experience" which unlocks the beauty of social media in sharing stories of great customer service travel far and wide.

Collaborate with Your Customers

I don't care whether you own a community or your rent one from a provider like LinkedIn Groups—a community is a great place to create a social experience with your customers. B2B marketers often host bi-annual advisory boards or user group meetings and these are just screaming for a way to keep them connected via a tool like an online community. A recent Jive report of over 2000 customers showed outstanding results. Lower costs, more revenue—a perfect combination!

- Users
  - Generated 32 percent more ideas
  - Sent 27 percent less email
  - Found answers to questions 32 percent faster
- Employees
  - Spent 42 percent more time communicating with customers, which led to better retention rate
- Results
  - Support calls dropped 28 percent (Lower costs!)
  - Sales to new customers jumped 27 percent (More revenue!)

In summary, social media consists of a powerful set of tools that yields strong bottom-line results for your company, but you need to take a holistic approach and not treat it like just another broadcast media.

10 Make Social Media a Listening Platform

By Dennis Shiao,
Director of Product Marketing, INXPO

In any conversation, whether it's a sales pitch, a staff meeting, or a chat over coffee, I find it more productive to listen, rather than speak.

As product marketers, we're well equipped to communicate. We craft messaging, position products, author white papers, deploy marketing campaigns and tweet to our heart's content. However, to be effective in product marketing, we need to listen as much as we speak. In fact, in any conversation, whether it's a sales pitch, a staff meeting, or a chat over coffee, I find it more productive to listen, rather than speak.

Why is that? Because by listening, I'm able to better understand the people "on the other side of the table," listen to their thoughts, and digest what they're telling me. In a conversation, I can then process this information "on the fly" and sustain a productive exchange. Productive conversations and meetings simply cannot occur in the absence of listening.

In this era of constant connectivity and the "24 hour news cycle," the social web has enabled an endless amount of status updates, wall posts, blog posts, and online chatter. While part of our job as product marketers is to be a part of this "micropublishing" environment, we also need to shift from our fingers to our ears. Instead of typing on the keyboard, we need to open our ears, far and wide.

We can start with the social media channels we're managing as product marketers. These channels include Twitter, Facebook, LinkedIn, and blogs, to mention a few. Set up searches for your company name and your company's products, as well as key terms in your market. Also, be sure to search for mentions of your competitors and their products.

Next, look to the growing marketplace of social media listening tools. There are a wealth of offerings on the market, some for free, some available under a "freemium" licensing model, and some you can license at a higher investment level. These tools can automate the listening for you, and alert you to mentions and conversations on the social web that you need to know about. For an example of social listening in the consumer market, look no further than Gatorade, which created an extensive Social Media Command Center in its Chicago headquarters.[iii] As a result of the social listening platform it employed, Gatorade says "it's been able to increase engagement with its product education (mostly video) by 250 percent and reduce its exit rate from 25 to 9 percent."

Social media listening platforms provide a low cost means to perform market research. You listen to what customers and prospects have to say about your market. And, by combining listening with a light amount of engagement, you can solicit input via social media, creating "virtual focus groups" online. By listening well, you understand the language that customers and prospects use to describe your products and services. From there, you can adapt your messaging and collateral and you can use the intelligence gained to optimize your Search Engine Marketing (SEM) campaigns.

Now that you've dedicated time for social media listening, you'll be better prepared for your outbound communications. Your tweets on Twitter, posts to your Facebook brand page, and postings to your blog are now optimized to speak in your customers' language. Your communications also cater to the business challenges that your customers and prospects face. In fact, your "editorial calendar" (e.g. white papers, blog postings, videos, etc.) can and should be driven from intelligence gathered via your social media listening platform.

Create this "virtuous cycle" and your customers win. And that means that you win as well.

11 Leverage Social Media or Be Fired!

By Sandra Greefkes, Vice President and Principal Consultant, Filigree Consulting

Product marketers focus on the customer, not just the technology. People want to talk about people, problems, process, and solutions, not products.

"Social media is like having a child for the first time. You don't know what you don't know."

Social media is more important to product marketers than anyone else in an organization. Yes, that is right, social media is important for corporate communications, public relations, and marketing communications; but, given social media's potential for timely and actionable insight, it is the product marketer's closest lifeline to guaranteed success in the market. Why? Because it gives you direct one-on-one access to customers!

As a product marketer you want to become a social media guru. Find out who in your company is responsible for social media and work with them to become engaged. Your product managers, customer service support team, and others will thank you. Most importantly, your clients and prospects will seek you out.

There are four social media actions you can take as a product marketer: listen, learn, share, and cultivate.

Listen: If you do nothing else with social media, you should at least listen. Not just once or from time to time, but continuously. Set up alerts, use an automated monitoring system, find the most relevant blogs, and set up an RSS feed.

Learn: What is being said, where, and by whom? Is there agreement or are there conflicting points of view? Who are your detractors? Who are your advocates? Actively seek out feedback. Social media allows fast, direct, and broad feedback that is cost effective. Anything from product specific feedback, to new product or feature ideas, new target segments, support requirements, etc. Join or start the conversation. Think of what you learn from social media as an adjunct to your in-depth research such as focus groups and surveys.

Share: This is not about how many followers or fans you have. It's about knowledge sharing with your customers and prospects which is supported by broad collection of thought leadership content—not corporate, product, or service information.

Cultivate: Enable customer intimacy and trust, through reciprocal sharing of valuable knowledge between you and your prospects. Form a virtuous cycle to drive deeper loyalty and better products and support. It's about enabling internal and external education and then applying that to your messaging.

Product Marketers are a social media marketer's dream partner. They have deep market knowledge and understand the issues that clients and prospects face in a market. They have deep solution knowledge which enables them to answer questions, react on the fly, and generally be as helpful as possible in the social media world. **Product marketers should focus on the customer, not just the technology.** People want to talk about people, problems, process, and solutions, not products.

At the end of the day, social media strategists and product marketers are a marriage made in heaven. You could not find a more synergistic relationship if you tried. So, if you are a product marketer looking to become a superstar, seek out and engage the people who lead the social media strategy in your organization. Otherwise you eventually might find yourself in the unemployment line.

12 Reinvent Word of Mouth Marketing

By Nick Coster, Co-founder, Brainmates

Word of mouth communication can be so effective that people can quite often discard other information that might indicate competing messages.

There are a variety of statistics that demonstrate the difference and impacts between positive and negative customer experiences. Happy customers will tell three people about their experiences, while unhappy customers tell more than ten.

We call this effect "word of mouth" communication.

The power of word of mouth communication is that most people will rely more on recommendations from people that they know than from strangers. It leverages the levels of trust that people share that are often difficult to establish.

Word of mouth communication can be so effective that people can quite often discard other information that might indicate competing messages.

Product marketers have always sought after positive word of mouth by sharing testimonials and other positive experiences that seem to come from a trusted source to overcome the inherent trust barrier that most people have when they initiate a relationship with a new person or brand.

The last ten years have seen the power of word of mouth change significantly as the Internet has become increasingly bi-directional. The rise of Web 2.0 allowed customers to post comments on websites. Blogs have allowed anyone to easily publish their ideas and share them with an increasing audience.

Yet it is has been the introduction of social networking services that have replicated and reinforced the historical trust relationships that people have always shared. Along the way social networking has allowed both weak and strong relationships to remain persistent—relationships that would otherwise fade away as people's social circles change and evolve.

Before social networks, if you lost contact with a friend, you had to choose to let the relationship go or wait until a chance future meeting.

Now social networks hold these connections until they are deliberately severed. The act of 'un-friending' someone, while technically trivial, implies an almost malicious intent. A feedback loop that causes these social networks to grow is thus created.

Before social networks became popular, the effort required to share an idea or experience with our 'network' was significant. You would need to have their time and attention to share that message, so it had better be worth it! This is one of the reasons why bad experiences spread more effectively than positive ones. The bad experiences make better stories.

Now the effort of sharing an idea has been trivialised, with the most fleeting of thoughts shared to all our ever increasing network of 'friends' or followers who get to choose whether the idea is of value to them. To further amplify this effect, social networks provide simple yet powerful tools to share and propagate valuable ideas to their networks.

For product marketing, this means that the power of word of mouth has been amplified exponentially by social networks, so that good or bad messages about our products can ripple through the attention of hundreds or thousands of people in just hours. This can provide an opportunity for Product Marketing to inject ideas that are worth sharing into these networks, by:

- Monitoring positive comments from customers about offerings and sharing them.
- Providing regular communications about products and services without selling to develop a relationship that can create a following.
- Identifying negative sentiment when it first appears and addressing it effectively.

These can become turning points in a market's attitude to a product if a problem or communication is addressed effectively and promptly.

13

Emulate Twitter!

By Cindy F. Solomon *@cindyfsolomon*,
Founder, The Global Product
Management Talk *@prodmgmttalk*

The qualities that make Twitter seem inane and half-baked are what make it so powerful.
—Jonathan Zittrain, Harvard Law professor, Internet Expert *@zittrain*

In an Agile world, product marketers must employ the most efficient means available to engage in real time product/market conversations. Twitter is a powerful broadcast and content marketing vehicle for PMMs to master.

Twitter Qualities to Emulate

Concise: A tweet is limited to 140 characters. Be terse. Develop and share content that is precise intentional, informative, and interesting.

Open: Twitter is public and archived on the web. Be open to discovering the value of your product from others' perspectives. Dick Costolo, Twitter CEO *@dickc* says: "Pay careful attention to the things that people do with your technology/service/product, because some of them may have discovered a powerful use for it that has completely evaded you."

Cross-platform: Twitter works across all platforms, operating systems, and devices. Tweets are viewable from the Twitter website, mobile apps, SMS/text messages, Facebook, and LinkedIn status updates. Integrate product communications across all media to guarantee consistency and trustworthiness. Optimize content to exploit strengths of each platform. Target messaging to emphasize unique value for each audience.

Scalable: The ability of a system/network/process to grow quickly and/or accommodate and facilitate spikes in growth. Develop content that easily grows and extends reach, such as Frequently Asked Questions (FAQs) and glossaries. Be sure to include key words optimized for search engines. Incentivize your audience to add content with support forums, blog comments, and wikis. Encourage customer accolades and case studies. Repurpose and refresh existing content for on-demand reading. Utilize Twitter Chat hashtags to crowdsource new content and capture tweets into transcripts.

Tweeting Strategies to Implement

Listen & Monitor: Search Twitter for product-related terms, issues, events, and hashtags. Lurk at Twitter Chats and create twitter lists of your target audience and market space. Use tools such as Klout to identify thought leaders, prominent bloggers, and competitors. Then analyze their behavior and connections.

Create & Broadcast: Tweet valuable information worth re-tweeting at optimal times to promote product lines, provide incentives, invitations, and share industry news. For example: "Please retweet: [article name] [by @author] [shortened URL] [optional comment] [#hashtag]." Utilize URL shorteners (bit.ly, tinyurl.com, goo.gl) and attach keyword #hashtags.

Analyze & Track: Identify tweeting behavior patterns and determine metrics; study how timing and headlines affect retweets; label *"Favorites"* to save highlights and testimonials; monitor *"Follower count"* to measure distribution power; track the number of *"@Replies"* to measure community engagement; utilize social media tools that provide additional analytics.

Engage & Dialogue: Develop trust by being human, consistent, relevant, and unique as well as generous, attentive, and positive. Ask questions, agree, or challenge politely to trigger and extend specific conversations. Always acknowledge and reply (@twittername), retweet (RT) valuable information worth sharing, and thank people for retweeting. Use the Direct Message (@DM) capability to private message a follower (DMs won't be listed in public twitter stream). Host, sponsor, and participate in regularly scheduled #hashtag Twitter Chats, such as The Global Product Management Talk #prodmgmttalk to extend community vibrancy, increase followers, and deepen product/brand affinity.

14 Create Successful Launch Teams

By Jeff Drescher,
Founder, JCT Communications,
Previously Vice President of Marketing,
Pancetera and BakBone Software

Properly launching a product involves ensuring that each department within your own company knows about the forthcoming product launch and is prepared for it.

Once upon a time I was hired by a software manufacturer to implement the Product Marketing function. This was not a huge company, about $40 million in revenues and approximately 150 people; however, they were an established company with offices across the world and a product that was in its fifth generation. On my third day in the office, I received an email from the Director of Public Relations that was sent out to the entire company, with a copy of a press release that just hit the wire announcing a major new product version that had just gone GA that morning. Upon questioning, I found out this was the normal course of action with respect to how employees were notified of new product releases. I was completely floored that a company of this size and having a mature product had absolutely zero process in place with respect to launching new products.

One of Product Marketing's key roles within most organizations is the ownership of the product launch process. Properly launching a product not only involves creating the marketing collateral and messages that tell the world about the product—and creating a press release—but it also involves ensuring that each department within your own company knows about the forthcoming product launch and is prepared for it. A solid product launch involves the creation of a product launch plan, complete with a task list of deliverables, along with a launch team consisting of at least one member of each functional department across the company. The members of the launch team are responsible for their deliverables and for making sure that information about the product launch is disseminated throughout their departments—whether dealing with pricing, part numbers, internal systems, sales training, etc. PMM (or Project Management in larger organizations) is responsible for monitoring due dates to ensure the launch is staying on track.

I like to use the analogy of a product launch being like a "bow and arrow." You spend weeks preparing for the GA of your new product or version in launch meetings (pulling back the bow). And then just as you release the bow and the arrow flies toward the target, you reach your GA date and all of the hard work preparing for launch pays off with maximum exposure for your product or company. The key to a successful launch is having everyone across the company on the same page with the launch and having all of the deliverables ready by the time the product goes GA and is announced to the market.

This means having the PR person sending out the press release that is set to hit the wire tomorrow morning, sales already trained on the product, analysts briefed, your channel partners up to speed, part numbers already in place through your distributors and internally, your website ready with the new content, etc. After all, your company spends hundreds of thousands, even millions, of dollars creating this product. Doesn't it deserve the best chance possible to be successful?

15 Marketing Checklists Ensure Success

By Jenny Feng, CMO, Marketeers Club

In the absence of a solid process, relying on memory can be costly. A checklist is a safety net.

The genesis of checklists can be traced back to 1935. A small crowd watched the Boeing Model 299 speed down the tarmac, lift off briefly, then suddenly stall, turn on one wing and crash in a fiery explosion. The investigation found "Pilot Error" as the cause. No one was more qualified than Major Ployer Hill to fly the test plane. However Hill was unfamiliar with the aircraft, and had neglected to release the elevator lock prior to take off. His mistake cost three lives and Boeing lost most of the U.S. Army aircraft contract to its competitors. This incident resulted in the first pilot's checklist—a simple approach to making sure that nothing is forgotten during takeoff, flight, landing, and taxiing. With the checklist, other pilots managed to fly 1.8 million miles without another accident. Variations of the checklist are still in use today.

In a different industry, Peter Pronovost, a medical director at the Center for Innovation in Quality Patient Care at John Hopkins Medical Center in Baltimore created a checklist comprised of five simple interventions to eliminate catheter-related bloodstream infections (CR-BSI) in the ICU. His checklist prevented forty-three infections and eight deaths, and saved $2 million in cost.[iv]

So, if checklists save lives in aviation and in the ICU, imagine what it could do for your business.

A successful product marketing manager is either armed with an MBA degree, years of experience or in most cases, both. She/he has the innate ability to effectively multi-task, organize, and attend to a myriad of details. Nonetheless, even the most competent and experienced managers make mistakes or skip important steps under pressure to meet multiple deadlines. In the absence of a solid process, relying on memory can be costly.

A checklist is a safety net that expedites what we heard in Rule 14. It is a method to ensure that the simple stuff gets done right the very first time. MarketeersClub.com recently conducted an interview with more than a dozen product marketing managers. One of our questions was: What would you like to change in your daily job? The top three responses were:

- Get things done in less time.
- Have more time for creative thinking and long-term planning.
- Acquire a larger budget so I can do more to grow my business.

At the end of the day, it's about reducing the lead time it takes to complete projects (speed to market) and reducing the cost to maximize profit.

Checklists may be created in many areas of a Product Marketing Manager's role and responsibilities—from pricing to distributions and promotions. As an example, a marketing promotion checklist might include the following:

- Define campaign objective and target audience.
- Assign campaign codes to track redemption.
- Make sure the promotion program complies with State and Federal Laws.
- Provide operations with forecast/anticipated unit lift from the promotion efforts so they ramp up production and build inventory.

Continuing on with the promotion example, think back on your last few promotional campaigns:

- How long did it take to create the campaign?
- How many rounds of revisions does it typically take to "get it right?"
- What have been some of the costly mistakes in real dollars (pre or post campaign)?

When creating a checklist for a product promotion, or any other product marketing tasks, consider the following three points:

- Start a list with what you already know (this could be a team or individual effort).
- Add to the list as you complete additional projects.
- Make a commitment to utilize the checklist every time.

The bottom line is: invest the time to create checklists customized to your category needs. The few minutes it takes to refer to your checklist will ensure that your projects are executed effectively and in a timely matter. You will get things done more quickly, leave your office earlier, and still be at the top of your game.

16 Remember Your Internal Customers

By Jennifer Berkley Jackson,
Founder, The Insight Advantage

The broader view you can take of who your customers are, the more successful you and your products will be.

It can be overwhelming if you stop to really think about all of the people in your organization who you need to keep in mind when launching new products. It goes way beyond the customers who are in your target market.

Below is a rundown of some of the people who rely on you to keep their needs in mind:

Sales Team - You need to not only provide great messaging for your sales team so they can sell the heck out of your products, but you also need to make sure that you understand what type of tools/materials are most helpful to your sales team and deliver what they will use. Creating cool printed brochures could be a total waste of money if that isn't the type of collateral your salespeople find valuable. Be sure to talk to them about what is most helpful in their efforts. Shadowing them on sales calls can be extremely illuminating...and help you make sure that what you do is relevant/useful.

As noted in the bonus rules, if you have a marketing or sales automation system, ask your salespeople which marketing assets they have learned to monitor closely when a lead has enough points to advance it to become a sales referral. Likewise, work with sales management to confirm marketing campaigns are creating qualified leads. If not, they can help you make sure that campaigns are targeting the right audience and that the messaging is setting up expectations that can be fulfilled, as well as share competitive insights that may provide you with key differentiators.

Support Team - Regardless of what your product/service is, someone will be supporting customers during installation and afterwards. Sometimes that is a formal tech support team, but in other situations, it's a customer care group that takes orders, handles returns, answers questions, etc. You can make their jobs easier by keeping them well informed during the product launch and by syncing up with them to develop supporting collateral that addresses the most common user questions. It's also critical to provide the support team with sufficient training so they are ready when the product launches. Plus, the support group is a valuable resource for real-world insights on how your customers are using your product that could be used for testimonials.

Finance Team - Your Finance team is extremely interested in the decisions you make about product pricing policies, discounting, etc. If that's part of your PMM charter, it's imperative that you work with them to thoroughly understand profit targets and all of the cost elements associated with your product/service to ensure that the product can be profitable.

And more... Operations, Engineering, Product Management, Marketing Communications, Web Team, and Executives are all stakeholders in the decisions you make about positioning and promoting a new product. And don't forget that your job of meeting their needs isn't done once the product is launched. Be sure to check in periodically with each of your internal customers to see how things are going and to see what needs to be fine-tuned. The easier it is to integrate your products/services into the organization, the more likely they are to have long-term market success.

The broader view you take of who your customers are, the more successful you *and* your products will be. As a PMM, I found that the effort that I put into making other people's jobs easier relative to marketing my products was greatly appreciated and led to invitations to work on many highly visible cross-functional teams, which increased my exposure within the organization and definitely helped my reputation as a valuable, strategic employee. A checklist (Rule 15) really helps!

Who is missing from YOUR list of internal customers?

Wield Influence, Even When You Have Little Power

By Reena Kapoor,
Founder & President of Conifer Consulting

Bring the customer's voice to the product strategy table and you will never have to worry about power in the organization again.

One of the challenges product marketers face is that you are sometimes asked to market a product which you had no say in defining, or worse, you may not see exceptional customer value in. The pig shows up at your door, so to speak, for some lipstick and no matter what you do, you know this pig may not fly! So what's a smart and diligent product marketer to do?

The key here is to find ways to influence the product development process and outcome long before the pig shows up at your door. In my career, including my consulting practice, I've observed successful product marketers—like yourself—who do a few things really well. Not only do these practices make your work more meaningful, it helps you wield influence when you have little power and it also makes you invaluable to the organization.

Be the Voice for Your Target Customer: Even as product strategy is being developed, bring your knowledge to those discussions. Sometimes the business is looking for a market (need) to address. In such cases, product marketing has an important role to play in identifying which markets are large, growing, and addressable with your business assets and satisfy other criteria. As a product marketer you should be integral to leading this effort.

On the other hand, if you're in a situation where your company has a cool technology looking for a market, once again, you should be helping/leading the effort to identify the target market, your target user (often different vs. target buyer if you're in a B2B business), and your core promise to them. If you bring the target customer's voice to the table, then your influence on product strategy, definition, and the future road map will not just be welcomed—it will be sought out. The key here is to make sure you're seen as a thought leader in seeking, developing, and integrating such customer knowledge with the company's or division's objectives. The best product marketers understand their customers *and* freely share this knowledge to strongly influence the initial product strategy and roadmap. Bring the customer's voice to the product strategy table and you will never have to worry about power in the organization again.

Zero in on Value and Differentiation: Before you develop one piece of marketing collateral, ask all key stakeholders two critical questions:

* What's the VALUE your offering will create for your target customer
* What's DIFFERENT about your offering vs. other offerings/substitutes that your target customer can use

Why would they pick, try, and stay with your offering? Get your team/organization focused laser-like on these aspects. And lead the team to agree on one or two sentence answers for these questions—not a whole paragraph where your poor pig tries to be all things to all people. Good product marketing managers make sure their product strategy teams remember these as they develop product requirements—and as YOU develop market positioning.

Help Product Managers Make Smarter Choices: The best product marketers also use their customer understanding to help product management make smarter choices about features and function enhancements. This can be done in many ways and I leave the politics and diplomacy up to you; but the idea here is that product managers often have to make trade-offs and tough choices for new product releases and they always appreciate a customer and criteria-driven approach vs. a shoot-from-the-hip one. They may be forced into the latter in the absence of good data or lack of time to gather good data. This is where you, the well-informed, communicative, focused product marketer can weigh in and influence the selection of features/ functions for enhancement releases.

Customer and market insights about what your customers need—both articulated and unarticulated—where the market is headed, and what your competition is up to are invaluable to this discussion and to your product's success.

18

Use Online Metrics for Product Marketing Success

By Dan Olsen, Product Management Consultant, Olsen Solutions LLC

Using these online metrics will help you measure your product marketing efforts and make them more successful.

Products are increasingly delivered and marketed online. Tracking a product's online metrics has grown in importance and has become a critical competency that separates leading companies from their competitors. This rule shows how to combine online metrics with a high-level product marketing framework to gain a better understanding of your business and how you can improve it. The framework analyzes the "per customer" economics for the three key parts of the customer lifecycle: acquisition, conversion and retention.

Acquisition means getting a prospective customer to visit your website or landing page. Prospects can arrive at your website through organic (free) traffic or through traffic that results from your advertising and marketing campaigns. The important metric for acquisition is *Cost Per Acquisition*, or *CPA* for short. To compute CPA for a given time period, you need to know how much you spent on advertising and marketing campaigns in that period. Divide the total amount you spent by the number of prospective customers that came to your website from your ads to determine CPA. Your CPA will vary by marketing channel, so it's best to calculate CPA for each channel so you can compare them. Google Analytics lets you track how many visitors your website had and where they came from. If you are using Google AdWords, it will calculate the CPA for each campaign for you.

Conversion means converting prospects to revenue-generating customers (e.g., getting them to buy your product). The top metric for conversion is your *conversion rate*. The conversion rate for a given time period is the number of new revenue-generating customers divided by the number of prospects and is expressed as a percentage. The cost of acquiring and converting a new revenue-generating customer is your CPA divided by your conversion rate. Obviously, lower CPAs and higher conversion rates are better. Google Analytics and Google AdWords let you easily set up conversion tracking. Conversion rates are affected by the messaging you use on your landing page and how it's designed. A/B testing is a common technique used to try out several variations of a landing page and statistically determine which performs best. Google Website Optimizer is a free tool that automates this process.

Retention means keeping your revenue-generating customers (e.g., getting recurring revenue or follow-on sales as appropriate to your business model). An important retention metric is your *retention rate*: what percentage of customers you keep from one period to the next. Your retention rate will determine another key metric: your average customer lifetime (in months or years). Another important metric is the *average revenue per customer (or user)*, often shortened to *ARPU*. To calculate ARPU for a particular time period, divide total revenue by the number of customers for that time period. Using your gross profit margin (your profit margin before marketing expenses), you can use these metrics to calculate the *Lifetime Value (LTV)* of a customer. The total revenue your average customer generates is your ARPU multiplied by your average customer lifetime. To calculate LTV, just multiply that total revenue figure by your gross profit margin:

Lifetime Value = ARPU × Average customer lifetime × Gross profit margin

Now let's tie acquisition, conversion, and retention together. In a healthy, profitable business, you want to make more money from each new customer than it costs you to acquire and convert that customer. That difference is your *average profit per customer.*

Average profit per customer = Lifetime Value - CPA/Conversion Rate

Your CPA and conversion rate will vary by marketing channel (and LTV can, too). So the goal is to experiment with new channels, see what values you get for your key metrics, and plug them into the above equations. You'll want to spend more on the best channels and prune the under-performing channels.

Using these online metrics will help you measure your product marketing efforts and make them more successful.

19 Apply SEO Fundamentals Everywhere

By Eric Krock, Founder, AIDSvideos.org

When creating content for the web, you are writing for two audiences: the search engines that index the content and the people who read it.

If you've neglected to study SEO techniques or failed to consistently apply them, you're in good company. When Bill Gates launched his personal blog, it had very basic SEO mistakes.[v]

When creating content for the web, you are writing for two audiences: the search engines that index the content and the people who read it. Unless you please the search engines, you won't reach the people. To ensure maximum reach for your content and maximum benefit for your organization, know and apply the fundamentals of search engine optimization (SEO)!

Identify your target keywords and key phrases. You want your blog post, page, video, or slide deck to be returned in a high search result position when users search for these terms. Use the free Google Keyword Tool[vi] to determine which keywords users search for most frequently. Search for those keywords, and see which pages are returned and in what order. Study how your competitors have optimized their pages so you can be competitive. Consider targeting terms for which there is less competition and you have a better chance of ranking highly.

Use your target keywords repeatedly throughout your content.[vii] Put them in your: domain name, social media account names, blog, category, and tag names, web page and image file names, page title, H1-H3 elements,[viii] image file names and ALT text, first sentences of paragraphs, bullet items, strong and emphasized text, and the text of links between your own pages. Use the keywords consistently and repeatedly, but not so frequently that they annoy the human reader.

Create a good meta description for your page of 160 characters or less. Although it won't influence page rank directly, search engines may use this as the "snippet" describing your page. Good snippets increase clickthrough rates.[ix]

Put a blog post at a stable permalink URL.[x] Link to the post from your blog home page using its title as the link text and a handwritten summary to describe it.

Avoid having the same content at separate URLs within your site. If identical content is at two separate URLs within your site, the pages will divide link equity between them, harming page rank,[xi] and only one will appear within search results.[xii] If your site design requires that the same content appear at multiple URLs (such as for different navigation paths or query sort parameters), specify a "canonical URL" so search engines know which URL should be considered "preferred."[xiii]

If you syndicate your blog posts, ensure that syndicated copies include a link back to the original copy.[xiv] Search engines will use this when consolidating search results.

If you put your trademark in your blog or web page title, put it at the end.[xv] For example, use "Road Runner Trap 2.0 Released | Acme Widgets," not "Acme Widgets: Road Runner Trap 2.0 Released." This keeps the SEO focus on your target keyphrase "Road Runner Trap."

Use the same techniques to SEO your online videos and slide decks. Use your target keywords in your video and slide deck titles, descriptions, tags, playlist titles, playlist descriptions, playlist tags, and account profile. Include links back to your website in your descriptions and profile.

By applying these techniques, AIDSvideos.org attained over 2.5 million views for HIV/AIDS prevention education videos with no paid promotion. Organic views are earned views. Apply the SEO fundamentals and maximize your reach and impact!

20 Generate Demand, Not Leads

By Christine Heckart,
Chief Marketing Officer, NetApp

Before you agree to be measured by leads or pipeline contribution, do the math!

Marketing and sales often use "lead generation" and "demand generation" interchangeably. These terms overlap, but point to very different mindsets and results. *Leads* identify buyers at the right point of the purchase cycle so you can step in and make the sale while *demand* takes a longer viewpoint by influencing buyers so that consideration of your offering begins before they start to actively search for a solution. Both activities make the sales team more productive.

Leads vs. Demand

Here's an analogy illustrating the big difference between generating leads and generating demand:

Say you're a teenager who wants to go to a party Friday night. On Wednesday, you call everyone you know to see if they're having a party, but learn either there's no party planned, or you're not on the invitation list. That random, dialing-for-dollars approach is what lead generation usually is like. Now let's take the analogy from a demand generation perspective.

You're the same teenager who wants to go to a party Friday night, but now spends time talking to people who might someday have a party. You nurture relationships who understand you are fun to be around. When they plan a party, they call you to make sure you're available because your presence is that important to them.

In a business context, lead generation can be anything from buying a list and spamming them with emails, to capturing prospects at trade shows or when they download a white paper, browse your website, or respond to an e-mail campaign. The more targeted the approach, the higher quality the lead. But alone, lead generation is insufficient because it isn't generating demand.

Demand generation entails talking early and often about who you are, what you do, why you're better, and when to consider your offering. It means being part of an ongoing conversation, and shaping that conversation in terms you can win. You'll know you've generated demand when prospects contact you at the first articulation of a problem.

Quantity vs. Quality

Every buyer's journey encompasses multiple touch points. Some are easy to measure—which is fabulous—but many are difficult or impossible to quantify. Usually it's the combination of all the touches, not any one element, that creates demand and ultimately results in a sale.

I understand lead-generation's appeal, because it's measureable; however, not all leads are created equal. Lead generation often devolves into a numbers game. If you ever find a marketing organization generating enough leads to satisfy its channels—or a sales channel happy with the leads it gets—call *The Guinness Book of World Records*. You'll be the first!

Before you agree to be measured by leads or pipeline contribution, *do the math!* Your industry will have a standard ratio of how many leads you need to pour into your sales funnel to generate a certain revenue flow out of the funnel. If you have enough budget to create a *material* impact on that revenue outflow, then own the objective! If you don't, then rethink how to leverage your budget for the biggest impact on creating and accelerating sales opportunities.

Generate Demand, Not Leads

Generating demand encompasses many components, including a qualified and interested buyer, but it is not limited to lead generation, and leads are just one way to achieve the revenue goal. For most businesses, quality matters far more than quantity. Never lose sight of this. My point is not to purge lead generation from your marketing toolbox, but prevent you from counting leads as your primary measure of marketing success.

A dollar spent in marketing should make every sales person more effective and efficient and result in customers knowing when to consider your solution, having a basic understanding of your differentiated value, and demonstrating preference for your brand. This is demand generation.

Orienting your marketing efforts to generate demand will create sales leverage, a bigger pipeline, faster close times, and ultimately more revenue, plus some high-quality leads!

21

Help Your Sales Team Communicate Your Message

By Tom Evans,
Principal Consultant, Lûcrum Marketing

Product Marketing Managers must own the message and make sure all market facing employees understand the message and know how to communicate it.

In my 25+ years of business experience in both sales and product marketing roles, I have experienced too many instances in which the sales team was provided some marketing collateral and a product presentation. Then people wished them luck as they prospected and tried to close deals with anyone who would listen. While this approach might be sufficient for the star sales people, the other 80 percent of the sales team pursues opportunities that don't fit well with your solution, speak to prospects that aren't really decision makers, sell solutions that you don't really have, and the list can go on. The overall impact is wasted time and effort in pursuing the wrong opportunities, confusion in the market place on what the company does, poor sales results, and eventually market failure.

Product Marketing Managers must own the message, *but owning the message isn't just about creating it.* It is also about making sure all market facing employees understand the message and know how to communicate it. Most important among these are the sales people and for that reason, I call this "Sales Enablement."

The most important goals of a strong sales enablement program are to ensure that:

- The sales team clearly understands the target markets and the target buyers within those markets.
- The sales team understands the problems, challenges, goals (pains) of the target buyer and can communicate a clear and consistent message on how the solution can address those needs.

- This understanding is shared by all market facing team members (sales, marketing, executives, etc.).

I recommend the following core set of enablement tools:

Product Backgrounder - This is used as a reference for the sales person to quickly understand the key aspects of the solution to help guide them in their sales pursuits and to easily discuss the product at a high level. At a minimum, the product backgrounder needs to include a succinct product description, definition of target markets, market challenges your product address, typical buyers and their main pain points, and finally your market messages (e.g., positioning statement, value proposition, etc.). You should aim for about one page and no more than two pages.

Needs Discovery Grid - This tool helps a salesperson carry on an intelligent conversation with a buyer and help the buyer admit to their pains and needs. You need to develop one for each buyer profile in each target segment.

Executive Level Presentation - Once you have uncovered the needs of the executive level buyer, this is a high level discussion of ten to fifteen slides that explains to the buyer that you can help them solve their problems and reinforce the key messages and benefits of your solution.

Message Driven Demo - Too many demos are about showing features and don't really show how the solution solves the buyers' pain points. A message driven demo tells a story that demonstrates how the solution addresses these pains and reinforces the key market messages.

There are, of course, additional tools and information that you can create for sales enablement, but you should deliver at least these four core documents and keep them concise enough that the sales team will actually use them.

Remember, don't just hand these tools to the sales team and hope they use them. You must proactively train them on these materials and then test them to make sure they adequately comprehend the information. And not to be redundant, but you must also train the other market facing team members previously mentioned. By implementing a sales enablement program as described, you significantly increase the potential for success in your target market.

22 Give Sales the Right Selling Tools

By David Fradin, Vice President
and Senior Principal Consultant, 280 Group

The sales force is your front line army. They get shot at every day and that constant rejection is not easy.

Your job as a Product Marketing Manager is to make sure the sales force has all the training and weapons they will need to be successful. You first need to understand some basic principles of sales forces and channels:

- Sales people follow the path of least resistance. They understandably prefer to spend their time on easier deals rather than pursuing complex, difficult opportunities.
- Sales people follow the money. If management tells you, as the PMM, to drive $10 million of your product's revenue this year but the sales force's compensation plan (commission and quota) doesn't include your product, you will not meet that revenue target.
- If you give your sales people unqualified leads, they will find out the prospect was just interested in the T-Shirt or iPad giveaway and had no real interest in your product. The next time you give leads to that sales person, they simply won't use them.
- People buy from the people they trust.

Give your sales force and channel the tools they need for success. The sales force is your front line army. They get shot at every day and that constant rejection is not easy. If they don't have everything they need to sell effectively, then they will be running back to you (making you into a "product janitor," as we discuss in the 280 Group's Optimal Product Management training) all the time to help with each sale. Thus you become a tactical operator instead of the strategic marketer you should be. Or your product will falter and somebody might notice that you are the one responsible.

The sales tools listed below need to be in available in print, in electronic format, and on the website.

Selling Tools Aimed at the Customer:

- Introductory paragraph about the product, which the sales person will use as their oral pitch ("elevator pitch") and in the cover letters (emails) with attachments. It needs to be short and punchy while positioning the product and motivating the prospective customer to want to learn more. Cover the key selling points derived by the unique selling proposition from the work the product manager did on product positioning, competitive research, and market research.
- Description of the product, its benefits, features, advantages, specifications. This is usually in the form of a brochure, sales sheet, or data sheet. This information is derived from the work product management did on business case, market requirements, product requirements, and competitive research including personas, product features, advantages and benefits.
- Draft proposal including terms and conditions.
- For trust and credibility building: customer collateral, depending on the type of product and customer, typically includes white papers and presentations, backgrounders, references, and quotes. If the product will save money, give the Return on Investment calculation or provide an ROI calculator.
- Sales presentation, covering all of the points in the bullet above.

Selling Tools Aimed at the Sales Person:

- Description of the qualifications and/or characteristics of the target customer based on work done on user personas, market research, and market segmentation.
- Why they should sell the product. How selling the product will be to the advantage of the sales person.
- Sales training presentation covering the above.

Selling Tools Aimed at the Channel (Distributor, Retailer, etc.) Principals:

- Description of the qualifications and/or characteristics of the target channel (distributor, retailer, etc.) based on from work done in market research, competitive research, and market segmentation.
- Why they should sell the product. How selling the product will be to the advantage of the channel.
- Contact for each level of the distribution channel as appropriate.
- Sales training presentation covering the above points.

With the right sales tools, sales people can focus on their goal: to make sales.

23 Successful Channel Plans Start with Strategy

By Mara Krieps, Founder and Principal,
Pivotal Product Management

Success in channel planning requires near-perfect alignment with company strategy.

Successful channel planning consists of managing distribution channel partners and developing channel marketing programs. In working with both large and small product organizations, I've found that novice Product Marketing Managers (PMMs) tend to view the scope of channel management as a series of promotional programs to be planned and executed. But seasoned product marketers start with the strategy, even if their work is focused on promotions. Here's what they know that you should too!

You'll have a far greater chance of promoting a successful channel plan when you start with the company strategy, clarify the product strategy, and align the channel plan with the other four Ps of the marketing mix.

Start with the Company Strategy

Success in channel planning requires near-perfect alignment with company strategy. For example, if your company is following a "price leadership" strategy and you are creating a distribution channel for a new product, you'll select channel partners based on how operationally efficient they are, or you may choose to sell direct in order to keep costs down. On the other hand, your company may pursue an exclusive, "first to market" strategy. If so, and assuming your product strategy follows the company strategy, your channel partners may need to be hand-picked (or a direct sales team), and your channel marketing programs need to reinforce that cachet.

So your first step is to get acquainted with your company's strategy, whether stated clearly in a document or implied through budget allocation and action.

Clarify the Product Strategy

Product strategy usually follows company strategy. But sometimes it doesn't. Think Starbucks "Via"—an instant coffee product from the purveyor of fresh brewed coffee and coffee beans.

Before you start channel planning you'll want to clarify how much your product strategy aligns with the company strategy and what direction your product strategy follows ("product differentiation," "cost leadership," or other), so you can be very clear about how the channel plan will support them.

Continuing the Starbucks example, if you were the PMM for "Via," which channel strategy would you promote: selling through all existing Starbucks stores and distributors, or through only a subset of them, or selling through a whole new group of distribution partners?

Align with the Other Four Ps.

The plan for distribution channel partners and channel marketing programs needs to line up with the traditional Four Ps:

Product features/functionality
- Is the product easy for channel partners to explain, sell, and support?
- What opportunities exist for channel partners to provide add-on services and customization?

Pricing
- What are the available margins to support the channel? Can partners make money?
- How many channel partners do we need to support the product revenue goal?

Promotion
- What ramp-up time do we realistically expect for channel partners? Have we built this into our revenue model?
- What budget is available to provide marketing programs (or marketing funding) to channel partners?

Positioning
- What is our competitive positioning?
- What channel structure would support our competitive positioning?

These questions are a good starting point for discussions with other cross-functional team members, including Product Management, Sales, Operations, Finance, Product Development, and Customer Support.

The successful Product Marketing Manager will get clear on the strategies, form a team, and begin these discussions.

24

Help Prospects Make Buying Decisions!

By Linda Gorchels,
Director of Marketing Talent Development,
Wisconsin School of Business

What words do customers use to convince themselves to make the decision to buy your product?

Do you like to be sold to?

I frequently ask product marketers that question. Most say *no*. So I probe deeper. "Have you ever been in a restaurant and asked the waiter what he would recommend? Or been trying on clothes and asked the associate for an opinion?"

If you answered *yes*, you were asking to be sold, but your perspective was, "please help me make a buying decision." So it is with your customers. They don't want a laundry list of features and benefits; they want decision-making guidance. And while the benefits of your product's features play a role in the decision, there are potentially a host of other factors as well.

There are emotional as well as rational motives for purchase decisions. Some decisions (such as for consumer goods) lean toward the emotional side while others (such as complex business products) lean toward the rational side. Both are important, however, since customers sometimes make an *emotional* decision that they *justify* using rational, logical factors. So arm them with the necessary tools to convince themselves *and* others that this is the correct decision. As you do this, remember that customers may choose to buy for reasons beyond your product entirely. Let's look at some factors influencing purchase decisions by asking a series of questions.

Why are customers making this decision? What are their goals?

Let's say buyers are making a decision to purchase lab equipment. Do they want ease of use, current functionality, or future capabilities?

Are they trying to position their company for efficiency or growth? Can corporate factors (such as brand, relationships, and history) influence the purchase decision? The more closely the product marketer can align with the customer's goals, the easier it will be to facilitate a buying decision in your favor. Demonstrate how your product can contribute to their profitability, efficiency, or competitive advantage.

What words do customers use to *convince* themselves to make the decision to buy your product?

Here is an example of almost verbatim promotional material from one of my clients, modified slightly to camouflage the company and industry. *"XYZ Company's solution is a comprehensive, integrated, and strategic customer care solution consisting of products and services that provide analytical capabilities, channel integration, process and sales improvement, and subject matter expertise to the industry."*

Say what? Do customers really talk that way? What is the mental self-talk they engage in to become more comfortable with their decision? Think about talking with customers and revise the statement in the prior paragraph. What "analytical capabilities" are important to the target customers and how does that relate to their goals? What is the importance of channel integration or subject matter expertise? And how should the prospect *feel* about the purchase (relief, trust, premier status)? What information do they need to believe this is the lowest-risk decision for them (testimonials, statistics)? If you can't answer these questions, customers will be thinking, "So what? Who cares?" That makes it really hard to help them make the decision to buy your particular brand.

How (and where) would customers prefer to buy the product?
While many product marketers might not have authority over this decision, it can nevertheless be a critical component of the purchase process. Different market segments have different expectations regarding online purchasing, immediate availability, access to complementary products from other manufacturers, or related services such as installation and repair. Simply focusing on the benefits of specific product features ignores these factors. Build a case internally to be present where and how customers want to buy.

Product marketers will always need to balance science and art. A fresh perspective on *how* customers buy may yield new insights into improved marketing. Envision a conversation where you help customers make buying decisions by focusing on their goals, the preferred process for purchasing, and the mental framework of words they use to argue in favor of the decision.

25 Create the Right Messaging First

By Michael Cannon, CEO and Founder, Silver Bullet Group

What's typically absent from a product launch plan are the styles, categories, and types of messaging required for market success.

We have all read the research reports[xvi] from IDG, AMA, etc. that say over 50 percent of the content we produce is not relevant to customers, nor to the field/channel sales teams. It's exceedingly difficult to have a successful career, i.e., launch and manage successful products, with such a large ball and chain strapped around a product marketer's proverbial ankle.

A typical product launch plan usually contains deliverables such as a presentation deck, product brochure, white paper, application note, etc. That is needed, but what's often absent are the styles, categories, and types of messaging required for market success. The result, as the research indicates, is that most of the go-to-market content is product-centric/technical and descriptive. What's missing is content that is customer business objective-centric and persuasive.

The solution is to create the right messaging first, and then to develop your go-to-market content. Here is how to do it:

Establish messaging as a separate deliverable. Messaging is a summary answer to the buyer's primary and secondary buying questions, a.k.a. the key points that must be communicated in order to convince a person to engage/buy, and is integrated into content via the copywriting/creative process. Content, in turn, is the actual words you use, both written and oral, along with support visuals, to persuade a person to do business with your firm. Content can be delivered in the form of documents, audio, and video.

Determine the right styles, categories, and types of messaging needed for market success. The two primary messaging styles are descriptive and persuasive. The categories of messaging can include: Company, Solution, Platform, Product, and Sales (market segment) messaging. As an example, descriptive product messaging is the typical "what and how" content in a product brochure. It answers questions such as:

What does the product do?

How does it work?

What are the key benefits?

What features are included/optional?

Persuasive product messaging is the "why" content. It provides relevant, differentiated, provable, and understandable answers to the buyer's primary buying questions, a.k.a. persuasive messaging types, such as:

- "Why should I consider your product?" for demand creation
- "Why should I meet with you?" for meeting creation
- "Why should I change out my current solution for a new solution?" for opportunity creation
- "Why should I buy this new solution from your company instead of other competitors?" for order creation
- "Why should I buy now?" for urgency creation

These "why questions" are at the heart of every prospective customer conversation, be it online or off-line, that both Marketing and Sales must persuasively answer in order to convince a person to engage and buy.

Create and deploy your messaging. Once you have created your messaging, you're ready to deploy it into your go-to-market content, such as collateral, campaigns sales tools, and sales/channel support training.

Yes, it's more work and you already have too much to do. The question you have to ask yourself is: "Would I be more successful if I created less, more relevant and persuasive go-to-market content, using the ideas above?"

The collective answer from your peers is an unequivocal YES. What they found is that getting the messaging right is truly a silver bullet: it's the only deliverable that improves the effectiveness of all the go-to-market content. The typical impact is captured in comments like: "We increased our pipeline by over 20 percent and our win rate by over 10 percent," "We were able to take 15 percent market share from our biggest competitor," "I now have a lot more credibility with the sales teams and spend less time supporting them," "Almost overnight we were able to communicate significant competitive advantage."

26 Always Test Your Message

By Greg Cohen,
Senior Principal Consultant, 280 Group

Customer validation also applies to product marketing.

Experience is a harsh teacher. She has visited me many times in my career. This experience is how I learned that customer validation also applies to product marketing. In particular, it is equally important to test and confirm marketing materials and messages as it is to test product concepts and designs.

I was working for a small, educational publisher targeting science curriculum for the elementary school market. Prior to development, we conducted a number of interviews with teachers. Two themes emerged:

1. Elementary school teachers lacked confidence in teaching science. The teachers were education experts, often possessing master's degrees. They were also subject matter generalists who were able to teach a single class about math, reading, writing, social studies, and history.

2. Elementary school teachers were universally pressed for time in their day, often finishing work up at home in the evenings. They were spread across many responsibilities with twenty to thirty students each. As a result, they often felt that they did not have enough time to do adequate lesson planning in science, which took them longer than other topics, because they were less comfortable with the subject.

My company used this research to create a powerful curriculum planning application to complement its science lessons. The teacher loaded the program on their computer and would check the national science standards that he or she wanted to cover in class. The application then assembled an entire lesson plan to match.

The company was very excited about this new teacher support application that complemented a well-designed science program. The company announced the new application with great pride at a small, education conference and trade show. The marketing materials prominently advertised that this science program, "Saved teachers time." It flopped! Reception to the announcement and the product was lukewarm. Something was wrong. Yet all the research said that, "Not having enough time in the day," was a major pain point for educators.

Not able to reconcile this point, I embarked on another round of interviews, this time to review the company's marketing. I gained great insight into our messaging with just a handful of discussions with elementary school educators. First, the message has to match where the customer is in the buying cycle. When considering a new program, the initial question for the teacher is whether the lesson as designed by the publisher will interest her students. Only if she believes the students will be interested, does she ask the second question: "Will I be able to implement this curriculum successfully in the classroom?"

Second, I learned that teachers found the marketing materials offensive. By promoting the main benefit of the program as being timesaving, teachers felt the company was suggesting that they were looking to take shortcuts. Therefore, not only was the company's marketing focused on the wrong message in the buying cycle, it also offended the very audience we were looking to persuade.

As a result, we changed the main message to "engaging students in science" which went to the heart of the primary buying motivation. We then focused the secondary message on ease of implementation, which conveyed a similar idea as timesaving without offending. The national level education tradeshow that followed the messaging change was a great success.

Fortunately for the company, the messaging error was picked up and corrected early. Nevertheless, I learned some valuable lessons, the most important being to test your marketing message with real customers. With the Internet, testing messaging is now easier than when I learned this lesson. For example, I can test the click through rates of different Google Adwords campaigns and the effectiveness of different landing pages without ever getting up from my desk. Nevertheless, even with this convenience, I still have to speak to customers to understand why one message works better than another does. Before I launch a product now, I always first test my message.

27 Forget about Your Product

By Bertrand Hazard,
Sr. Director of Business Strategy, SolarWinds

The best product marketers strive to live, breathe, and think like their customers.

"Business buyers don't buy your product. They buy your approach to solving their problem."[xvii]
—Jeff Ernst, Analyst, Forrester Research

Forget about your product.

Product marketing is less about *what* you sell and more about *who* purchases it. Understanding your customer is what matters most, mainly what they care about and *how* they purchase and consume products. Only at that point should you strategize how to position, package, price, and sell to your customers.

Planning with the end in mind (the purchasing transaction) distinguishes a good product marketer from a great one. Great product marketers map their marketing strategy and sales initiatives *around* their customers' buying journey. Most importantly, they strive to *live, breathe,* and *think* like their customers.

Doing so strengthens the product marketer's best effort to:

Create a memorable story. The best product marketers create an *emotional connection* with their audience by casting their customer as the story's main character and speaking in their customer's voice. They search diligently for the perfect tone of authenticity that resonates best with their customers and focus their message on solutions, benefits, and value without superlatives and buzz words. Think Apple's iPod *"1,000 songs in your pocket"* or Southwest Airlines *"You are now free to move about the country."*

Create an enchanting experience. The best product marketers lead their customers through a beautifully orchestrated buying experience. They appreciate that their customer cares about the *cumulative product purchasing experience* including the initial website visit, first interaction with the sales team, online registration to community and support sites, and so on. They continue to escort their customers throughout the implementation phase ensuring that each customer is getting the *promised* value from their product. They know, especially in the B2B space, that a cohesive purchasing and on boarding experience will build the strongest *relationship* with their customer and increase the likelihood of an initial and subsequent purchase. Think Rackspace's *"Fanatical Support"* brand promise and delivery.

Create passionate brand advocates. The best product marketers transform enchanted customers into their most effective sales agents by nurturing their customers' passion and providing them the communication channels to voice their love for the product, whether an online forum or offline user conference. They root for their customers more than anyone else because they know that in these days of growing peer influence on purchases, their best customers are also their best marketing assets, their best brand advocates and ultimately, their best sales reps. Think Harley Davidson and their *"Live to ride"* official riding club (HOG) or Zappos and their mission to *"Deliver Happiness."*

With the growing use of social media, product marketers have more ways than ever to connect, engage, and insert themselves into their customers' habits while building a genuine relationship with their customers. With *intimate* customer knowledge, product marketers are best positioned to lead their company's go-to-market strategy and execute upon Peter Drucker's vision:

"The aim of marketing is to know and understand the customer so well that the product or service fits him and sells itself." [xviii]

Every journey begins with a single step, so slip into your customers' shoes, walk a few miles in their problems, and...*forget about your product.*

28 Help Customers Cost-Justify Your Product

By Rich Mironov,
Managing Director, Mironov Consulting

An ROI calculator should show current costs and how your product/ service will directly save the customer money.

The market is full of customers who **might** be interested in what you're offering, but don't have the time or energy or focus to push themselves through the buying cycle. One place where they often stumble is in justifying the money they'll spend on your solution. In most B2B markets, individual buyers have an approval process that demands some kind of justification, often with multiple levels of signoffs.

So important questions for Product Marketing are *"How will the customer justify paying for your product?"* and *"What tools are you providing to make this decision easier?"*

Easy to ask, but often hard to answer in practice. Vendors are typically much more motivated to make a sale than customers are to buy, so you'll need to do some of the thinking for your intended buyer. That suggests a very simple spreadsheet or "ROI calculator" to capture the essential value of your story.

For a B2B product, the intended buyer isn't the CEO, but a line employee or manager. Sizeable purchases have to be explained / justified / "ROI'd." So you need to help your target customer explain the specific savings that come from your solution. Part of product marketing's sales enablement kit needs to be such a simple savings calculator. You should count only numbers and hard dollar impacts. Avoid "strategic value," vague improvement, and handwaving.

Starting from the customer's point of view, an ROI calculator should show current costs and how your product/service will directly save the customer money. Or, alternately, how you will help the customer make more money (top-line revenue). All of the logic and numbers are from the customer's side: their actual costs, quantities, numbers of transactions, which will be used to calculate the percent improvement, etc. Savings must be computed from as few inputs as possible, while still supporting your story. Neither the buyer nor the CFO wants to spend much brainpower on this exercise. For example:

- *"Our super-special credit scoring application will reduce the number of outside credit checks you have to run. You currently do {insert number} credit checks per year at {insert price}. We'll reduce that by {insert percentage} for a savings of {compute here}. We will only charge you {insert price} for a net savings of {compute dollars} and ROI of {percent}."*
- *"Our super-special cell phone gaskets protect cell phones from damage when your subscribers drop them into puddles or mugs of beer. You currently replace {percent} of subscriber cell phones every year for water damage, each of which costs you {dollars} in parts and support, shipping, and wasted time. Using our gaskets, you'll reduce this by {percent} or {total number per year} for a savings of {dollars}. Gaskets only add {our price per widget} to the phone cost, which is an overall savings of {compute dollars} and ROI of {percent}."*

And so on. The details vary, but the approach does not.

Consider a starter ROI template such as *http://is.gd/roisavings*:

- You will need to insert your own savings logic, similar to the examples above. Every savings story sounds similar, but the details are specific to your product/service.
- Make sure it's trivially easy to use, even for non-spreadsheet folks. This sample template has all inputs (entry items) marked in bold blue italics with colored borders. That makes it obvious for a user (customer or sales rep) to know which are the inputs. Likewise, all of the cells that are not inputs are 'protected' to avoid accidents.
- Keep things simple. If you can reduce the model down to only a handful of rows, it is more powerful. Make two (or more) versions if necessary. Don't have branching logic or super-fancy math that you'll be endlessly explaining to sales reps who aren't spreadsheet gurus.

Now go close some deals!

29 Sell More by Talking Less

By Joe Pulizzi, Junta42 Blog
Founder, Content Marketing Institute

To be a content marketer we must first realize that our customers don't care about us, our products, or our services. They only care about themselves.

Traditionally, to sell more product, we needed to talk about our products and services as much as possible. We needed to do more advertising. We needed to position the product message wherever our customers were located.

Those were the days when our customers had fewer choices for information. According to Forrester Research, 90 percent of buyers begin their search for a product or service on the web. Customers can now get practically all their information in the form of informational blog posts, reviews, social media referrals, eBooks and more. To that point, Google's Eric Schmidt stated that every two days we create as much information as we did from the dawn of time until 2003. Amazing.

Am I going to tell you to create more content? Well, yes.

Content marketing, one of the fastest growing marketing areas according to the Content Marketing Institute (CMI), is the idea that all companies are publishers. To attract attention, we need to be creating valuable, compelling, and helpful information, just like a media company does.

Eighty-eight percent of all companies employ some form of content marketing today, but most aren't seeing it result in more customers. Why? The biggest reason, according to CMI's study, is the challenge to create content that is truly valuable.

Let's face it, although publishing has been around since the dawn of time in some form, this is a new skill set for marketers. Marketing professionals are so used to talking about themselves, it's very difficult to develop content strategies that **don't talk about their products or services.**

To be a content marketer we must first realize that our customers don't care about us, our products, or our services. They only care about themselves. So how do we attract customers and prospects that only care about themselves? Create the best content on the planet in our niche. Here's how to get started:

Set Up Listening Posts: The first step to content marketing is to listen just like a publisher or journalist. Talk directly to customers and open feedback channels with customer service and sales. Be sure to use customer surveys and analyze keyword search terms using Google's Keyword Tool[xix] and look for story ideas using Google Alerts or Twitter Search.

Log Your Customer Questions: Log every question your customers have ever asked you. By doing this, you are mapping out your story ideas and editorial calendar.

Create Your Publishing Hub: The content needs to be located somewhere. WordPress, Hubspot, or Compendium are three solid options for your content platform. Think of this as your content library.

Focus on Content Distribution: Now that you've developed your content hub, start to think about the different ways to distribute that content to customers.

Use tools like Twitter, Facebook, LinkedIn, and Google+ as well as email/print newsletters, guest blogging/webinars, and eBooks/white papers.

Develop Your Calls to Action: More than anything, we want to make a connection with the reader in some way. These could be small or big, such as liking you on Facebook, subscribing to your enewsletter, downloading your eBook or white paper, trying a demo and, of course, buying something.

Keep the Editorial Calendar and Repurpose: If you don't set a content schedule, the work won't get done. It's as simple as that. Once that is complete, look for ways to repurpose your content into other avenues that would be valuable to customers.

Results? This strategy takes time, but if done well, you can be found everywhere your customers are searching for information. We've been at it almost five years now, and built a multi-million dollar business from nothing. Now, everywhere our customers search online to find answers to their problems, we are there, and this drives our business.

Done right, it can drive your business as well.

Additional resources are available.[xx]

30 You Are Not Your Customer

By Alyssa Dver, CEO, Mint Green Marketing

If you can't convey your product and its benefits to customers in two sentences or less, something is wrong.

"Customer" can mean the **buyer** (person who signs the check for the purchase), the **user** (person(s) who actually use the product), and the **influencers** (people who have a say in whether the company buys it or not). You can never be all those roles for whatever product or service that you sell.

Whether your product/service is business-to-business or consumer-to-business, there are almost always multiple customers in each sale and knowing them is paramount. 'Knowing' includes their demographics, their buying cycles, their budget requirements, their price sensitivity, their experience with similar offerings, etc. You can only know these things by asking. No amount of web research or analysts' reports are going to help with that. You need to do primary research—even if on a small scale and done informally.

I often get asked to do primary research on behalf of clients who are too busy, too intimidated, or too fearful of what they may hear from prospective clients. The outsider (e.g. consultant) can ask the naive and politically incorrect questions while the respondent can feel a little less worried about honestly answering questions. That middle layer often exposes more truth than the vendor will get directly. To get the most insightful information, talk to the people who decided not to purchase your product/service. It's best to do such research early and often.

Your Mother Actually Does Need to Understand What You Do

Don't assume writers, analysts, or other "thought leaders," let alone your target customers, understand your product. Even with industry experience, chances are we all have our own baggage and unique perspective that may or may not help us know what your product/service does.

To know if your positioning and sales message works, try it on mom or the neighbors and even your kids. If they can understand it well enough to paraphrase it back to you, you've succeeded. Don't assume it's too complex for them because if it is, most likely it's not the product but your words that are complex.

You can't put words in people's mouths nor ignite word-of-mouth unless you control your own.

If you can't convey your product and its benefits to customers in two sentences or less, something is wrong. I am not suggesting that you can explain your entire offering or competitive value in two sentences, but a good positioning that makes it clear who, what, and why separates the targets from the others quickly. Most of us are fearful that we may offend or reject potential friends. In marketing, filtering the market is actually good! If you can separate wheat from chaff, targets from uninterested, you don't waste sales time and marketing resources. Ideally, if your positioning is clear and specific for those who care, those who don't will politely say, "That's not for me." Good, then both of you can move on.

Once you have the target person intrigued, you do need to provide more information to convince them that your product is not only right for them, but the best one for them and something they absolutely must buy (ideally now!). Doing this should be done in incremental, meaningful steps. That is, providing a twenty-page datasheet may be too much for the buyer to invest at the early stage of the buying cycle. However, offering digestible amounts of information to the buyer is really important so they continue to seek more and are willing to invest their time knowing your offering seems to be a good solution for something they are needing. Slamming them with a folder full of datasheets, brochures, article reprints, case studies, and press releases probably is a bit overwhelming to most any customer in any business. The same applies to your website.

Marketing isn't about convincing someone to buy something they don't perceive they need but rather effectively educating about what you have to offer them and why it's of clear value to them.

31 Prepare Your Marketing for the Final Three Feet

By John Cook, Disruptive Products Guy, Nokia

In a Business to Consumer (B2C) environment, we often have to leave the task of selling the product to the customer to those who may be underskilled, unprepared, or both.

As product marketing managers, we spend a lot of time thinking about positioning our products. However, in a Business to Consumer (B2C) environment, we often have to leave the task of selling the product to the customer to those who may be underskilled, unprepared, or both. That is where The Last Three Feet concept really counts. When selling through retail stores or distributor and consultants, you must give as much attention to crossing the last three feet of delivering your product as you give to the miles that it traveled before it reached the end seller.

We're not selling lemonade in front of Mom's house anymore. We often don't sell the products we craft. That final sales task is frequently left to those who can't highlight the right features, are unable to address customer concerns, or sometimes, may have no idea of what the product's features and benefits actually are. In fairness, we must recognize that retail associates, sales teams and the like, are often expected to be "experts" in a huge range of products. We cannot expect them to know everything.

So, what's a top-notch product marketer to do?

First, consider **your selling channel** as your target customer that can help that final interaction with the end-user. If you don't sell into your channel first, (and well), you can lose the opportunity to sell to your end-user at that critical final three feet. As PMM's, we drive the product's sales cycle: consumer product consideration, sales, and support.

Product consideration and sales support are the bookends that support the all-important selling process. Your customer, the channel, needs to be armed with the facts to help the end-users get themselves through the consideration phase.

If your sales associates feel your company provides strong marketing support, they'll be more confident to better answer those sticky, but critical, customer queries and do it effectively.

Consider these three questions:

- Do you spend time fine-tuning the web copy and data sheets from both an end user and channel sales perspective? Consider the audience carefully.
- What demo materials have you created that a channel partner can leverage? Will the demos work for an untrained channel sales person? Are they easily accessible in a busy retail environment?
- Who's training your channel support team and major retailers? Shouldn't you be involved in that?

In my years of B2C and B2B experience, these activities rank as high as the classic four P's of marketing: product, price, place, and promotion because they are the critical last step, when the retail store clerk has to explain the product to the end customer.

A PMM team I once led took the channel-as-customer concept to the next level. It was as simple as integrating ourselves into the selling process. Along with our channel managers and sales teams, we briefed, (under NDA), channel partners many months before our products were due to come to market. Yes, we risked someone leaking details, but the reward was that we gained insightful feedback about everything that affected our mutual businesses at in-store placement opportunities. After we did this a few times with one major retailer, (at both their headquarters and ours), we actually ended up co-funding a research project. We now had valuable insights into our targeted customers' buying preferences in a way we could have never obtained by ourselves. The risk paid off.

Your distributor, your direct sales channel, your retailer *are* your customers too. Don't forget or neglect them and you'll close that last three feet with more of your end-users, be they large corporations or your next-door neighbor.

32 Protect Your Product's Rear

By Jim Anderson,
President, Blue Elephant Consulting

If you are successful, you're going to have competitors trying to find ways to take your product marketing crown from you. Will you be ready?

So there you are, your company has launched its fantastic new blue widget product and the reviews are rolling in: you've got a hit product on your hands. Life is going to be good forever—or is it? If you are successful, you're going to have competitors trying to find ways to take your product marketing crown from you. Will you be ready?

The product marketers at everyone's favorite on-every-corner coffee store, Starbucks, were faced with this very problem.

McDonald's, Dunkin Doughnuts, and just about every other fast food chain out there started to offer espresso-based coffee products. Oh, and they did it at prices that were lower than Starbucks. The Starbucks product marketers saw this happening, but they were constrained from a direct competitive response because by necessity Starbucks has a limited menu (in order to keep things moving quickly) and the majority of their business is done in the morning. Great, so what was a coffee product marketer to do?

Seven years ago Starbucks bought one of their competitors: Seattle's Best Coffee. They hadn't done that much with them before then, but thanks to the new competitive threat, the Starbucks product marketers decided to change all that.

Starbucks came up with a plan to roll out a second coffee brand called, you guessed it, Seattle's Best Coffee. This is the brand that they decided to use in order to compete with the lower-priced fast food coffee offerings that were nipping at their heels.

This is where the real product marketing innovation occurred: the new brand was to be sold everywhere. That means that it started to show up in about 30,000 different locations ranging from fast-food outlets, supermarkets, bookstores, and even vending machines.

I can almost hear some of you starting to mumble concerns about this approach. You've got a good point. Since Starbucks was to be selling most of the Seattle's Best Coffee through franchisees they were going to have a significant quality control issue. Additionally, the coffee that is sold though vending machines is generally the worst coffee out there. What was Starbucks thinking?

The product marketers at Starbucks realized that they really didn't have all that much to risk. Starbucks had less than 4 percent of the U.S. market for brewed coffee. There was no place to go but up. Additionally, the engineers at Seattle's Best created a new type of vending machine that they believed could actually deliver a good cup of coffee.

Starbucks found itself in a difficult spot that all of us product marketing types wish that we could find ourselves in: it had the #1 selling coffee product. However, the competition was coming on strong. The product marketers at Starbucks realized that if the competition was successful at taking away the lower end of their customer base, then there will be nothing to stop them from moving up and capturing more and more customers over time.

Starbucks took the novel approach of launching another brand: Seattle's Best Coffee. They hoped that by distributing this product everywhere they would be able to take some of the wind out of the sails of their competition (and money out of their pockets).

For product marketers everywhere there is a lesson in this for all of us. Your competition will always be looking for ways to take market share away from your brand. Under the right circumstances, developing a second brand can give you a way to battle the competition on their home turf without any risk to your main brand. Now you have one more product marketing strategy to be successful with!

33 Be the Customer Usage Expert

By Dave Kellogg, Technology Executive and Blogger, former CMO, Business Objects

What will change your career is becoming the expert in how your customers use your product.

There are plenty of people in your company's engineering and product teams who are experts in how your products work. Over time, those people are typically seen as valuable resources for the company (as in, "No one knows more about the optimizer than Joe").

Product Expert	Usage Expert
Talks about technology	Talks about applications
Justifies technology value	Justifies business value
Understands how features work	Understands why people need features
Knows current competitors	Knows where the market is going
Seen internally as valuable resource	Seen internally as organizational leader

If, however, you're looking to both impact revenues and be seen as an organizational leader, then you must become the expert not on how your products work, but on how your customers use them. Here's how:

Engage Constantly with Customers

Thanks to social media, it's never been easier to engage with your customers. Use a blog, Twitter, Facebook, and LinkedIn to enable direct web-based customer communication. Make yourself easy to find on the web and easy to contact. But don't stop there.

Run periodic and topical surveys so you can not only watch needles move over time, but also stay on top of the market pulse. Attend industry conferences and user-group meetings. Give presentations so people can find you. Most importantly, work with your sales channels to set up as many live customer meetings as possible.

Ask the Basic Questions

In customer meetings, learn to say: "I don't know." Once you start pretending to understand a customer's business, you're sunk. Start with "I'm sorry, I don't know anything about medical products distribution and it will help me greatly to have a basic understanding of your business." Then pepper the customer with why, how, and impact questions.

- Why do your customers buy from you?
- How does our product help you make money?
- What's the **impact** of failing to meet the service-level agreement?

Watch Your Language

Remember that you are delegated to the level at which you speak. If you start talking bits/bytes or saying "orthogonal" too much, you're likely to find yourself talking to someone in a cubicle outside the corner office. Marketers must be bi-lingual: speak tech to techies and business to business people. Mix the two at your peril.

Develop Legendary References

As you meet customers and learn the business impact of your products, invariably a few people and a few stories will stand out. The stories will be easy to understand and impactful. The people will have an unusual passion and willingness to tell them. Embrace these few people and turn them into legendary references. Feature them in case studies. Invite them to speak at user groups. Place them at industry conferences. Connect them to the business and technology press. Ensure they build relationships with your top executives. Teach the entire company their stories and how your product affected their businesses.

My favorite legendary reference went from being an internally focused CIO to a repeat InformationWeek 500 award winner and right-hand to his CEO. They'd cross the country on the corporate jet telling their customers how my product helped them provide the best service in his industry.

Do that. And then watch what happens to your career when people say: "Nobody knows more about our customers" than you.

34 Speak in the Customer's Language

By Jeff Lash, Creator of the Q-and-A site "How To Be A Good Product Manager" (http://ask.goodproductmanager.com)

The best product marketing connects with the customer by speaking in terms which they understand and which resonate with their needs.

Have you ever told family members or friends that you work in marketing and get responses of chuckles and eye rolls? Have they followed up with jokes about "paradigm" and "synergy" and "world class"? Their reaction is understandable, given the sheer amount of bad product marketing that is out there in the world. Bad marketing gives marketers a bad name, but good marketing can improve the impression of marketers by increasing overall marketing effectiveness and driving companies' success and revenues.

Speaking in the customers' language means two things: making sure that your language and approach resonates with the market, and making sure that you are tailoring your marketing effectively to the type of buyer/persona.

Too often websites, brochures, and other marketing materials are filled with jargon that only makes sense to people within the company or to industry insiders. Sure, your whole product development team might know what the acronym means, but does the customer? Does a "scalable, best-of-breed cloud-based framework leveraging cutting-edge XYZ technology" mean anything to your buyer (or even your sales team, for that matter)?

Customers don't care about features, they care about benefits. They don't care about whether your solution is "built on the newest JRE platform" or not, so long as it "works on all devices." Saying that your product "addresses key government regulations" is much clearer—and more concise—than listing a handful of acronyms.

Not only do you need to talk about benefits in language the reader will understand, you also need to think about the target reader(s). For most products—especially those in business-to-business markets—there is not a single individual "customer," and you need to address those buyer personas differently. With an enterprise IT purchase, for example, the IT manager deciding which product fits the organization's needs the best is going to have different questions, concerns, and lingo than the finance manager approving the expenditure. In that case, the IT buyer may really want to know about the technical jargon, while the finance buyer may be more concerned about long-term maintenance costs and licensing terms.

When you address the right buyer in their terminology, it allows them to determine better whether the product addresses their needs and allows you to overcome any potential hurdles they may have in the purchase process. There are few things worse than convincing several key stakeholders that you have the right solution for them, only to be hung up on one required approver because your messaging didn't speak their language.

The most ineffective product marketing is filled with internal jargon, overused hyperbole, and targets the wrong audience. The best product marketing connects with the customer by speaking in terms which they understand and which resonate with their needs.

By understanding the language your buyers use, and understanding the differences between your different buyer personas, you'll not only make your marketing much more effective, but your company revenues may grow dramatically. Plus, as a side benefit, you just may change the reaction you get from family and friends.

35 Create Simple Messages for Complex Products

By Noël Adams, President, Clearworks

Remember this: customers don't want details and technological explanations, they want benefits and functionality.

All too often companies have great products, but those products never have a chance of being seen by customers. Why? Many times, customers don't understand enough about what the product is to even want to check the product out. You can have a great product, but even a great product will fail without the right message.

Companies with great products often fall into the trap of wanting to tell the customer everything. Product marketers are so excited about their product that they want to capture all of the attributes in one explanation. With the help of their internal product teams, they get caught up in the details and wind up including even the most technical information in their product message, somehow afraid that they might miss the chance to convey something that one customer might find important. The result is typically an overwhelming, confusing message.

Remember this: customers don't want details and technological explanations, they want benefits and functionality. They want to simply and easily understand your offer and how it fits into their world.

How many times have we all looked at a website for a product and thought "wow—I have no idea what they are selling." The message is just too complex. Messaging should be clear and simple even for the most complex products. Complexity doesn't just refer to the jargon and technology terms used, it also refers to the excessive detail that gets in the way of communicating the key points. The key is to strip out that complexity.

We have clients who come to us when they don't understand why a product is underperforming. When we do the analysis, we often uncover that the problem is not the product itself, but the product message. People may have a need for the company's offer but didn't understand that the company actually offered what they want. How do you fix this? Rework the message. The first step in rewriting the message is to engage the customer and understand what is important. Talk to customers about what they look for in a product like yours. Listen for the key points that are important to them and equally, what is unimportant. Listen for the words they use to describe the product and the benefits they espouse so that you can speak in terms that make sense to a customer, not a product marketer. Next, write your message in simple terms, include only those points that are important and use terminology the customer understands and can relate to. And finally, test that message with customers. Make sure people get what you are saying quickly and easily without further explanation.

To create simple messaging, remember these key points:

- Focus on what the customer wants, needs, and cares about
- Use words the customer understands and can relate to easily
- Save the technical details for the appendix
- Test the message with your target market; let them tell you if they understand and find it compelling

In the future, if you have a product that is underperforming, look at the message. And remember, the litmus test for simple and clear messaging is not whether your internal team can understand the message, it is whether the customer understands, relates to, and finds the message compelling. Go talk to customers and simplify your message.

Turn Your Audience into Advocates

By Adrienne Tan,
Director, Brainmates, Australian product
management and marketing agency

In effect, social media becomes the platform where users sell to buyers.

A key task of Product Marketing is to create buying messages that entice buyers to purchase. Sound simple? As any great Product Marketer knows, there is a plethora of activity that occurs before crafting appropriate messages that make products stand out in crowded markets, some of which include:

- Market segmentation
- Eliciting buyer problems
- Developing a buyer persona
- Mapping the buying process
- Understanding the product, its features and benefits

Product Marketing is a significant investment for any business. To get it wrong means that substantial efforts across the business may not be realized. On the flip side, getting it right means that the rewards are plentiful. The investment in Product Marketing and the cost of doing business means that we must seek ways to stretch our limited resources. Using the target audience as advocates for the company's product and message is one way in which we can effectively do more with less.

Social media has made it possible to generate enthusiasm and excitement amongst our audience and subsequently create a more natural, believable way of selling our message. In effect social media becomes the platform where users sell to buyers. Peer recommendation and endorsement are extremely powerful, more so than stale ads.

How do we turn our target audience into advocates?

Deliver a product that solves a problem better than the competitors. While Product Marketers are not directly involved in Product Development efforts, outwardly focused views on the market and knowledge of the buyers means that Product Marketers can influence the final product solution.

In order to turn the audience into advocates, the organization must provide a talking point. Product experiences that delight the customer from the beginning to the end are the only meaningful talking points that drive significant business value. Without this, the audience will certainly talk within their circle of influences, but the conversation may not be positive and may damage the product's reputation and brand. Good product experiences are talked about.

There are some simple rules to follow.

Be where your customers are.
Product Marketers who stay on top of what their customers are saying and doing are ones that will know how to create loyal advocates. In order to do so, you'll need to find your customers and be a part of their community. Alternatively you can create an appropriate community for your customers. In order to establish a loyal following who will spread your product's key messages, do both. Join in and create a community.

Engage and deliver value.
When you've found your customers, best to sit back and listen first. No one likes a noisy marketer. During this stage, identify influencers of the community and trending topics. Once you're familiar with the community, genuinely engage with your audience. Over time, you'll discover key opportunities for offering your 'tuned in' customers a little more 'love' so that they start to spread the message and become your advocates. Needless to say, what you offer to your customers needs to be of value before an exchange occurs.

Make it easy for your audience to be advocates.
If you've created a community, make sure it's easy for your customers to participate. And this simply means providing the right tools, content, and messages for your customers to use and spread the good word.

Done genuinely, results will follow.

37 Get Close and Personal with Your Customers

By Natalie Yan-Chatonsky, Product Strategy Consultant, Brainmates

Product marketers can now engage more intimately with more customers than ever.

Talking regularly to both customers and non-customers is the basis for successful product marketing planning and execution. Conversations can easily be carried out one-to-one over the phone, face-to-face, or video conferencing, or one-to-many with social media. Product marketers can now engage more intimately with more customers than ever by eliciting deep insights about their profiles, behavior, lifestyle, and goals that relate to the product experience.

The value of developing a dialogue with target customers of your product experience is multi-pronged:

- Validate or correct any assumptions that the organization previously had about the target market.
- Gain new insights about the target market's behavior and unmet needs, which may have changed over time.
- Develop the value propositions for the product experience.
- Facilitate the development of messages that appeal to each target market.
- Demonstrate that your company genuinely cares about how it wants to serve customers better in the future.
- Build up loyalty with your brand.

The other key success factor to make customers care more about your product is maintaining an ongoing dialogue with your customers. Engage them to review the outputs of your marketing plan throughout the whole market planning process, even if the outputs are low-fidelity mockups. Iterative testing will reduce risk of developing messages that miss the mark with customers.

A Case Study:

A client was preparing their go-to-market strategy for an extremely complex product experience that many people within the company and industry struggled to describe in plain English without visual aids or tomes of text. They needed help to develop effective sales and marketing material.

After we built up a clear understanding of the organizational and product goals, our team carried out a series of customer interviews. We intentionally kept the structure for the interviews quite loose so that we were able to probe into any revealing comments. There were three factors of the interviews that enabled us to uncover what truly excited and frustrated the customers about the product experience we were researching:

- A sincere upfront explanation of how we would use the information about the customer.
- An empathetic tone to help customers open up to us.
- Ability to listen and read unspoken cues in the customer's tone and body-language.

Based on the insights we had gathered from the interviews, we were able to map out the customer problems, align the product features, and define the benefits for each segment and then to go back to the market to verify the priority order of both the customer problems and product features for each segment. This is where social media became useful: we could target higher volumes of product buyers and users within different market segments and frame our questions differently.

Once we had verified the priority features and benefits for each market segment, and had done further competitive market research, we created a relative comparison chart of how each feature compared with the competition's offering to develop the positioning statements and unique selling points.

Following an iterative approach while carrying out the marketing planning enabled us to develop the strongest product communications strategy that the client had ever had. The client's new salesforce was able to systematically follow a product sales process that pinpointed the exact messages for each type of customer. They also had the appropriate marketing collateral to provide at each critical decision making point to support their messages.

If product marketers regularly converse with their customers, their external communications will always capture the essence of what customers want to hear.

38 Honest, Open Messaging Leads to Success

By Paul Gray, Director of
Community Engagement, independent games
startup Bubble Gum Interactive

Increasingly, product marketers are recognizing that developing a valuable and beneficial relationship with customers requires a more open and somewhat humble approach.

Product marketers are responsible to make potential customers aware of products, encourage them to consider and use these products, and establish a loyal commitment to, or relationship with, the brand.

Despite this direction, our industry is littered with terms that suggest a somewhat combative nature. "Targeted marketing," "Clusters," "Campaigns," and "Cut-Through" all have practical meanings but do not help a profession that is sometimes viewed as underhanded or sneaky.

Increasingly product marketers are recognizing that developing a real relationship with customers requires a more open and somewhat humble approach. Just as your mother told you when as a child you may have broken the rules, it's always good to be honest.

Lying Can Be Powerful

We all know that marketing messages can have incredible effects. There is no shortage of studies showing how certain product marketing initiatives can increase awareness, stimulate interest, and create strong demand for products.

Yet numerous studies show significant and rapid declines in consumer trust of brands. Shifting customer attitudes, perceptions about corporate greed, and dissatisfaction with products or services in an increasingly competitive environment undermine some traditional means that product marketers have to persuade customers.

Today, consumers have even more power than ever. Social media has transferred ownership of brands from purely the domain of the brand itself to those that interact with it, including customers and critics. A product that does not live up to its promises may very quickly be attacked by customers who feel cheated or who are just simply tired of hearing "marketing spin."

Honesty Ultimately Wins

Product managers design, develop, and deliver the products and services that we as product marketers must take to our audience. Our colleagues create products our customers actually need or want and we must understand the underlying drivers, what problems or challenges the product solves, and how this fits into a competitive context.

Product marketers must accept any product limitations. It is no longer possible to get away with product features that aren't there or confusing them with complex positioning.

This type of deceptive marketing is common; consider a telecommunications carrier that provides multi-tiered calling rates at different times to different networks with $49 buying $300 worth of calls; a deal-of-the-day website that promotes discounted rates based off inflated prices that the partner vendor would never charge and so on.

The concept of money-back-guarantee is an example of this dodgy marketing. The definition of guarantee is that it is just that—guaranteed! That means whatever you promise *will* happen. To consider that it may not and be willing to offer money back is fine, but then it's not really much of a guarantee, is it?

Great products are built on trust. Product marketers must be honest and upfront in selling their product. This can sometimes involve explicitly acknowledging what a product is not.

The game that my company has created for children is purely an entertainment product, however we're often asked if it is educational. When we explain this is not how the product was designed, some people suggest we try to market it this way. There are very subtle educational components such as developing social skills, teamwork, problem solving, and financial literacy. This might indeed open up new opportunities for us, but we feel very strongly that this is unethical and misleading. We have built an entertainment experience and we will market it as such.

Establishing a reputation for honesty enhances product positioning and customer perceptions about the product and the wider brand. This only serves to improve the success of marketing and sales efforts. Sales teams know what attributes to focus on, customer service agents know how to manage expectations, and customers learn to trust and respect the brand.

39 Learn to Love Marketing Data

By Leslie Bixel,
Senior Principal Consultant, 280 Group

One small software firm was able to double their annual revenues without increasing their marketing budget or adding additional sales staff!

Bottom line: the success of a Product Marketing Manager will be determined by the revenue and profit performance of their product. The best Product Marketing Managers know that understanding performance data of marketing programs will aid them in driving sales and maximizing marketing ROI.

Before the era of online marketing, a Product Marketing Manager's marketing planning was largely dependent on faith in the value of image advertising and on historical data from other products, other strategies, and other customers, at best.

The Internet has provided "real customer, real-time" measurable forms of marketing. Digital marketing is traditional direct marketing on steroids. Faster, cheaper reach with instant feedback. This is fantastic! Now we can quickly demonstrate that product marketing makes money!

- Using Internet advertising you can get ads to targeted customers, and invite them to visit your website
- Using web analytics you can then track what these customers do on your website, and determine what and how they buy
- Using web optimization tools you can observe customer behavior and gather data how users interact with individual web pages
- And with good e-mail tracking systems you can measure the success of reaching and selling to customers receiving your e-mails

Does this mean a product marketing manager needs to become a web analytics guru? Of course not, but understanding what data to measure and track, and how to interpret the results, is a critical skill for every product marketing manager.

Small early stage companies often fail to capture the data that would help them tune their marketing efforts for maximum profitability. Larger, more established companies may have the opposite problem, and product marketing managers can find themselves drinking from a marketing data fire hose.

In both situations, it helps to put in place a simple spreadsheet report that captures a small set of performance metrics, and tracks them on a quarterly basis.

A summary report of important data trends provides the management team a quick and clear view of the impact of product marketing spending, and demonstrates the value of a product marketing manager as an orchestrator of business results.

Using free or low cost web marketing tools like Google AdWords, Google Analytics, MailChimp, Constant Contact, and others make it easy to produce a report on marketing performance indicators:

- Number of page visits, number of ad impressions, number of mail opens, and click-through rates
- Position of paid listings, organic search rankings
- Amount of increased web traffic (unique visitors), duration of website visits, ratio of new to returning visitors
- E-commerce information regarding the number of immediate sales generated for products sold on line, the length of the sales cycle, geographic trends, impact of sales promotions and pricing changes
- Number of leads generated for products sold online and/or offline. Increased volumes at call centers or retail outlets

Benchmarking and tracking performance this way provides the opportunity to experiment quickly, build on success, and eliminate wasteful spending. Using this simple tool with one small software firm recently allowed me to tune their marketing mix and improve individual program performance to such a degree that they were able to double their annual revenues without increasing their marketing budget or adding additional sales staff!

Learn to love online marketing data. Let real customer data feed your marketing strategy. Use trend indicators to maximize marketing ROI and raise your own worth as a contributor to the bottom line success of your company.

40 Shine the Light on Product Marketing

By Jennifer Doctor,
Managing Partner, HarborLight Partners

The problem is that even our executives do not truly understand what we do, and don't always understand why we do what we do!

Ever feel alone? Feel like product marketing is misunderstood? Feel your job is threatened? You're not alone, but you do need to take control and change it.

Product Marketing is different from Product Management. Product Management is responsible for getting the product on the shelf, while Product Marketing Management is responsible for getting the product *off* the shelf.

Earn Value by Working with Your Colleagues

When you are pushing through your tasks, are you respecting your internal partners? Do you work *with* marketing communications and others on their deadlines, instead of pushing your timing on them?

The basis of respect is trust and credibility. The quality of what you, how you, and when you are the basis for the credibility. How you communicate, to whom, and how often cannot be underestimated.

Talk the Talk *and* Walk the Walk

When you get a chance to be at the table, are you talking about your strategy and deliverables? Execs aren't interested in the amount of tactical effort and time you put into creating all those data sheets or webinars. These leaders are focused on results and business goals. Take the opportunity to talk about strategy.

- Talk about the product marketing roadmap you've created and how that is guiding you.

- Talk about your partnership with internal teams such as product management or sales.
- Talk about your contribution to the corporate strategy.

Keep this in mind: results drive the business, so align yourself with corporate goals.

Establish Your Own Distinctive Competence

You need to advocate for yourself as a value-add to the company, so start at the beginning. What do you do *better* than anyone else? What is unique about you?

Bottom line: if you don't have a unique quality or skill that adds a specific value to the team and differentiates you from the other team members, you are giving others the opportunity to define you—and maybe not in the way you want to be defined.

Engage in Communities!

In developing your distinctive competence, take a role in larger communities... not just your company. Engaging in product camps, twitter hashtag following, online communities, and the like that provide support to answer the questions you can't find within your company.

As you become more comfortable and involved, you will establish a voice. It's validation, growth, and leadership. It's a badge of honor you can show within your company. Step forward and develop a voice.

Remember that PMM is the only team/role/function concerned about how to communicate with our market at a holistic view.

Really? The only ones? Yes. Look at the primary roles of your peers:

- Product management is looking at the market problems.
- Development is looking at how to solve those problems.
- Marketing communications is looking at how to design the tools to present to the market.
- Marketing Services is trying to get people into the sales funnel.
- Sales is primarily concerned about the person in the market with whom they are speaking with at that moment.
- No one else is looking at how the market, those who have not already decided your product solves their problems, except product marketing.

Providing That Voice Is the Value of Product Marketing

In the end, your role as a product marketer is best developed by being the defender of your market, the superhero who represents the market voice at the leaders' table. When you do that in a focused and well thought out style with sound facts to support your claims and ideas, you will earn your value. And don't forget to always "Defend the Buyer."

41

Use Your Competitor's Products against Them

By Janey Wong,
Product Marketing Manager, Oberon Media

What's missing is the information on the specifics of what makes a competitor's product superior or inferior to your own in customers' minds. This information is the key to creating differentiation in your messaging.

Understanding the target customer is one of the most important tasks of a product marketer: the why, what, when, where, and how customers want to buy. But sometimes data is just not available about your target market, and with limited time and budget, typical research methods may fail to provide you with the insight you need. So under tight constraints I suggest you employ a quick and easy shortcut I use: bring in target users to experience your competitors' products.

The majority of research conducted—competitive analyses, benchmarks, and usability tests—tend to be primarily inward focused. These all provide very important information needed to help you position and differentiate your own offering, but they seldom venture into detail about the competitors' product offering and user experience. What's missing is the information on the specifics of what makes a competitor's product superior or inferior to your own in customers' minds. This information is the key to creating differentiation in your messaging.

Am I saying product marketers should focus precious time, money, and energy on gathering feedback solely about competitors' products? Absolutely.

By inviting your own customers in for a lab test, you can gather key insights about how your target market searches for and evaluates competitive products, what features help them inform purchase decisions, and how they interact with the products.

Since an opportunity to interview and spend face-to-face time with a customer is invaluable, it's easy to jam-pack a multitude of questions and exercises into one session. I'd suggest strictly keeping the research duration to less than one hour by prioritizing your research objectives to focus on uncovering the most difficult-to-obtain and inaccessible information about your competitors' products, and leaving sufficient time to hone in on the areas where you want the participants to provide more in-depth insight. For example, try to get to the root of why one competitor's product is better than another's to determine true customer needs and improve your positioning. A great method for drawing out what customers want is by using elements of Strategyn's Outcome-Driven Innovation technique that are pertinent to product marketing.

Lab tests can be intensive; they can also feel redundant to participants if you want them to compare your product against more than one of your competitors' products. To prevent this, design the sequence of tasks to be an interesting and positive experience for participants. They will be more focused on the tasks at hand. For example, to gather feedback on how you could best market new product features, you can brief your customers on the product enhancement goals and the rationale behind why you felt the enhancements were important for customers. This empowers participants to feel more like subject-matter experts, encouraging them to evaluate the products as critics and to give additional consideration to how the new product features would make a difference in their personal lives. Understanding your customers' thoughts about your product benefits will help you make your value statements stronger.

Warning: The risk to this type of testing is your customers might discover they prefer a competitive product. However, the value is that you'll know exactly why your customers might switch brands. It's the "why" that will help drive differentiation for your products against competitors in the marketplace, and assist in developing stronger key messages and go-to-market strategies.

42

These Are Our Rules. What Are Yours?

By Phil Burton, Senior Principal Consultant and Trainer, 280 Group LLC

We could have written 420 Rules of Product Marketing Management.

Rules are a way to guide future behavior and decisions, to minimize risks and maximize returns; or at least improve the odds of success.

Product marketing addresses a wide range of issues and challenges, as you can see by the range of topics and experiences in this book. Reading these rules, several key themes emerged. One of them is the importance of past experience.

Experience and wisdom often comes at great cost, as we saw in Rule 26. We were struck by the number of rules that were based on earlier mistakes. So we can think of these rules as a way of our contributors saying, "Look, we made a mistake, but we learned from that experience, and so can you, the reader." The railroad industry has a *General Code of Operating Rules* that addresses the issues of accident prevention in an inherently dangerous environment. There is a saying in that industry that, "All these rules were written in blood," which means they were learned the hard way, after a bad accident. It's not quite so dramatic for most of us, but we and our companies might be more successful if we take these rules to heart and develop a good set of best practices.

But even if we address all the issues that stand between us and complete success today, how useful would that collection of rules be? As this book is being written, we are all in the middle of a huge change wrought by social media. Many economies worldwide are either growing slowly or not at all. Without a doubt, the best practices have been changing and will continue to change. It's not enough to observe these changes. You need to keep asking, **"Why?"**

So what are you going to do? This closing rule is usually an invitation for you to consider your own rules. It almost goes without saying that you need to be on a lifelong quest for professional development, new rules, learning from your own experience and others' experience. Considering the range of issues, we feel that this book has just scratched the surface. We could have written **_420_** _Rules of Product Marketing Management._ As one example, a contributor commented that one topic alone, "Making the most of your marketing dollars," could be a whole 42 Rules Book of its own. As a practicing professional, begin to develop your own set of rules.

Let us know your thoughts. http://280group.com/blog/

A Bonus Rules

Bonus Rule 1: Unite Sales and Marketing to Drive Revenue

**By Gary Parker,
Product Marketing Manager, FalconStor**

Successful marketing companies are learning to transcend the old silos of sales and marketing to form teams that jointly manage both marketing and sales pipelines.

Ever heard hallway conversations like these?

Sales: "The only reason I didn't make quota is because the so-called leads from Marketing were totally useless and wasted my time."

Marketing: "We work twelve to thirteen hours a day to produce brilliant collateral with spot-on messaging that drives lots of leads, but they are all wasted on those people over in sales who couldn't sell free ice cream in the summer."

Both sides are each saying: "Your side of the boat is sinking!"

Research is showing that successful marketing companies are learning to transcend the old silos of sales and marketing to form teams that jointly manage both marketing and sales pipelines. It actually is not really "both" pipelines, but a single integrated revenue funnel that is monitored from a common view. That allows the team (which is now responsible for the entire pipeline) to make critical decisions on allocating resources. To accomplish that, they need a solid understanding of the costs for each stage from inquiry to closed deal with details by campaign so there is a common discussion point and decision criteria.

They also need to constantly clean their data to ensure they are working with accurate information.

As noted in Rule 3, as well as Steven Woods' book, *Revenue Engine*, understanding the prospect's digital body language and collaborative marketing sales/programs are critical to advancing a potential deal through the lead stages. As a good example of joint efforts to improve that process, the team can ask their sales people which marketing assets do you monitor closely when a lead has advanced to become a sales referral? Is there a particular white paper, webinar, eblast, web page visit, or other marketing asset that seems to be associated with hot deals? Are there any marketing assets that they have had bad experiences with? This valuable input is then used to adjust the lead scoring system to ensure that leads which flow from the lead nurturing funnel to the sales reps are optimized. It also strengthens your ability to personalize your marketing. This approach means that the sales team now helps own the lead scoring criteria instead of being just a recipient. Having an objectively defined score forms the foundation of a common view into the funnel.

Just as the sales reps provide feedback, the same thing happens when marketing management and sales management jointly own and review the revenue generation funnel. The team will monitor pipeline stages across web traffic / inquiry / marketing qualified lead / sales accepted lead / opportunity / closed deals to look for anomalies that can be corrected. For example, if the inquiry to marketing qualified lead ratio is low, then that might mean campaigns are targeting the wrong audience or the messaging is setting up expectations that cannot be fulfilled. A low marketing qualified lead to sales accepted lead ratio in one territory might actually mean too many leads are coming in or newer reps that need more training. The list goes on for insights that a non-politicized team approach will provide.

A company's survival depends on close collaboration between the marketing and sales groups. This not only helps drive increased revenue, but is also a much more pleasant way to spend the many hours that we all do in our marketing positions!

Bonus Rule 2: Keep Sales Messages Different from Product Documentation

By Phil Burton,
Senior Principal Consultant and Trainer,
280 Group

Effective marketing messages are structured around each step in the sales cycle and focused on the people who make the decision or recommendation.

I once worked for a startup company with a powerful application for network infrastructure management and configuration. This application combined three separate areas to automate labor and time intensive manual tasks. The benefits were more effective network management at lower cost. The product got a lot of buzz and won a "Best of Show" award. But the company never met sales goals, and soon shut down.

The product demo was very impressive but the key benefits were not apparent until about thirty minutes into the demo. And, selling any new product against established and well-known vendors with an installed base is a real challenge. The company had a compelling new product, but as a startup with no name recognition nor budget for a large awareness marketing campaign, the sales people did a lot of cold calling. In a first call, the salesman started with a fifteen to twenty-second elevator pitch, and hoped the prospect was interested enough to continue the phone call, and then agree to an in-person meeting.

We struggled endlessly to write a script for that first call, but could never distill the essence of that product demo down to fifteen to twenty seconds, about forty words. The problem came from trying to describe product features, not benefits, and basically not understanding the difference between product documentation and marketing assets.

Product documentation is aimed at users and people who support the product. The documentation explains product features, usage, and related information.

In contrast, the purpose of marketing collateral is to educate the prospect about the product, why it is different/better, or the benefits from buying the product or service, depending on the stage they are at.

To create effective marketing messages, you need to understand the buying process, or the sales cycle, for your product in the target market(s). The buying process is different for businesses and consumers, and it is different for big ticket items purchased infrequently vs. lower-cost items purchased on a regular basis.

You need to understand the steps and personas in the sales cycle starting with information gathering and leading through to the final decision. Who is the buyer? Are there key influencers who can recommend, or reject, the product? In a business, the user, buyer, and influencers usually have different job roles. In consumer markets, the buyer and the user can be an individual, or the household, and different family members can be the buyer, the user(s), and key influencers. The key issue for marketing messages is, "What is necessary to get the approval to move to the next step in the sales cycle?"

At each step short of the actual sale, the only measure of success is agreement from the customer to move to the next step in the sales cycle.

Effective marketing messages are structured around each step in the sales cycle, focused on the people who make the decision or recommendation. The messages need to address the specific information needs and concerns of those employees or family members. It is also important to understand how to deliver those messages, e.g. a face-to-face meeting, a web page, a data sheet or sales brochure, a TV or radio ad, a blog entry.

Please see Rule 3 on digital body language for more insights into tracking these stages while prospects are at the online inquiry stage.

Going back to the original example, the calling script should have focused on the product category and the key benefits, because "success" for that particular calling script was the customer willingness to continue the phone conversation, and continued customer interest with agreement to a face-to-face meeting. And so on, through the entire sales process leading up to an order.

B Contributing Authors

Noël Adams

Noël is president of Clearworks (http://www.clearworks.net) where she and her team of product experts help clients define and launch new products and services integrating customer insights and user research from idea through launch. The team has worked with large and small companies in both the consumer and B2B space. The Clearworks client list includes companies like T-Mobile, Dell, and Symantec in industries as varied as solar energy, wireless, financial services, and non-profit.

Jim Anderson

Dr. Jim Anderson is known as the "Real-World Product Manager" because he believes that product managers and product marketers can learn best by studying how other companies have accomplished what they are trying to do. As president of Blue Elephant Consulting, Dr. Anderson uses his practical, real-world experience to share his proven strategy for streamlining the development of products with clients. Dr. Anderson can be reached via email at jim@BlueElephantConsulting.com or online at http://www.BlueElephantConsulting.com.

John Armstrong

John is a Silicon Valley-based consultant providing marketing expertise in cloud, mobility, and security to technology vendors. Prior to his consulting career, John was VP and chief networking analyst for Gartner's global networking practice. He has held key management positions at Yipes Communications, Madge/Lannet, and

SynOptics/Bay Networks. John has a graduate degree from the Annenberg School of Communications at the University of Southern California, and a BA from Ryerson Polytechnic University in Toronto. Contact: armstrong@calcentral.com

Jennifer Berkley Jackson

Jennifer founded The Insight Advantage, a full-service research firm that provides a full range of services to help organizations gather and integrate customer insights into their businesses whether it be measuring customer satisfaction or helping prioritize features in a new product. As a consultant and prior product manager, she has extensive experience in successfully implementing business strategies across organizational functions, integrating the voice of the customer, streamlining and documenting processes to increase efficiencies and improve quality, and developing product strategies.

Leslie Bixel

With decades of product management and marketing experience at top ranked technology companies such as Adobe Systems, Leslie Bixel consults with senior managers on strategic marketing issues. As a Senior Principal Consultant at the 280 Group, Ms. Bixel drives initiatives to address common client challenges such as new product introductions, revenue growth, expansion into new geographic markets, international standards engagements, and corporate positioning. Ms. Bixel is a graduate of Stanford University and an alumna of the Harvard Business School.

Michael Cannon

Michael is an internationally renowned sales and marketing productivity expert and a best-selling author, most recently coauthoring with Jay Conrad Levinson (Guerrilla Marketing), et al., *Marketing Strategies That Really Work!* Michael is a CEO and Founder of the Silver Bullet Group and creator of the hugely successful Persuasive Messaging™ System. For more information, visit http://www.silverbulletgroup.com or call 925-930-9436.

Greg Cohen

Greg is a Senior Principal Consultant with the 280 Group and a fifteen-year Product Management veteran with extensive experience and knowledge of Agile development, a Certified Scrum Master, and former President of the Silicon Valley Product Management Association. He has worked and consulted to venture start-ups and large companies alike and has trained product managers throughout the world on Agile development, road mapping, feature prioritization, product innovation, product lifecycle process, and product management assessment. Greg is the author of the books *Agile Excellence for Product Managers* and *42 Rules of Product Management* as well as a speaker and frequent commentator on product management issues.

John Cook

John has been a consultant and executive in product marketing for almost twenty years. He has held senior level product management and product marketing positions in many high-tech companies including Apple, Palm, HP, and Nokia. John is a left brain/right brain person who's as comfortable talking with engineers as the customers who buy their products. He's a hands-on person, who's not afraid to explore and use the products he works with.

Nick Coster

Nick is a Co-founder of brainmates who is passionate about the benefits of putting the customer before the technology and building products and services that customers love. Nick has been developing and managing products for over thirteen years, with a diverse range of companies and industries.

He has developed a deep understanding of the way the different technologies fit together and is always amazed at the new and exciting ways that people use them. http://www.brainmates.com.au | *@nickcoster*

Jennifer Doctor

Jennifer is the managing partner of HarborLight Partners where she brings her nearly twenty years of experience in a unique blend of product management, product marketing, marketing, and IT knowledge to her work. A seasoned Product Camp veteran and supporter, she founded Product Camp Minnesota. You can read her thoughts on her blog at http://www.outsideinview.com; as a contributor to http://onproductmanagement.net; and via Twitter by following *@jidoctor.*

Jeff Drescher

Jeff has over twenty years' experience in Marketing, Corporate Strategy, and Product/Brand Management at major companies. Jeff led the peer-to-peer networking boom at Artisoft, Inc. managing the LANtastic product line and then helped shape the server-based backup market at ARCADA Software where he managed the Backup Exec product line. Jeff was Marketing VP at Pancetera and, before that, held similar positions at BakBone Software, VERITAS Software, Storactive, Compaq, and Artisoft. Jeff founded JCT Communications—a Public Relations and Marketing Services firm. (http://www.jctconsult.com)

Paul Dunay

Paul is Chief Marketing Officer of Networked Insights and a B2B marketing expert with over twenty years' success generating demand and creating buzz for leading technology, consumer products, financial services, and professional services organizations. He has authored four social marketing books and received numerous awards. His blog, Buzz Marketing for Technology, has been recognized twice as a Top

20 Marketing Blog. Networked Insights is the developer of a marketing decision platform that helps companies use real time data to optimize marketing.

Alyssa Dver

Alyssa is CEO of Mint Green Marketing (http://www.mintgreenmarketing.com) providing affordable, expert marketing consulting and training to large, brand name companies to up-and-coming start-ups based all over the world. Her books include: *No Time Marketing, Software Product Management Essentials*, and *Ms. Informed: Wake Up Wisdom for Women*. Alyssa is a media go-to with articles in *Forbes, BusinessWeek, Software Magazine*, and dozens more. See her at one of the many business or women's events she keynotes each month. alyssadver@mintgreenmarketing.com 508-881-5664

Tom Evans

Tom Evans is Principal Consultant at Lûcrum Marketing and is a recognized leader in the product management community. He has over twenty-five years of technology business experience, working in start-ups to Fortune 500 companies and has been responsible for establishing and implementing the product management and product marketing roles at multiple companies. Tom is a Certified Product Manager (CPM), Certified Product Marketing Manager (CPMM), and Agile Certified Product Manager (ACPM). Tom is passionate about creating and marketing winning products.

Jenny Feng

Jenny has held a number of marketing and product management positions in the consumer products industry. She led the growth and profitability of key business divisions at companies such as Staples, Avery Dennison, Delta Creative, and Behr. Jenny is currently the CEO of MarketeersClub.com, a leading resource for product marketing professionals, Managing Director at CorpMVP, Inc., a marketing consulting firm, and Adjunct Professor at Cal Poly University, Pomona. Jenny received her MBA from Babson College.

David Fradin

David is Vice president and Senior Principal Consultant for the 280 Group. He was a product manager and product marketing manager with Hewlett Packard Corporation, followed by product and business unit manager with Apple Computer. Subsequently, he worked as associate director of market research for the Personal Computer Industry Service at Dataquest. He is currently writing *Building Insanely Great Products*, the history and future of Product Management.

Linda Gorchels

Linda is director of marketing talent development on the Executive Education faculty, Wisconsin School of Business at University of Wisconsin-Madison, where she has trained more than 10,000 global corporate executives. An award-winning author, Gorchels has published the 4th edition of *The Product Manager's Handbook* (2011), *The Product Manager's Field Guide*, the forthcoming *Business Model Renewal* (2012), and co-authored *The Manager's Guide to Distribution Channels*. Prior to joining the university, she held numerous successful marketing positions with leading companies.
http://www.brainsnackscafe.com

Russ Gould

Russ, Director of Product Marketing at Guidance Software, has spent much of his career in technology product marketing creating go-to-market strategies, positioning solutions, and enabling Sales for success. The quest for product revenue growth led to a keen understanding of effective demand generation. While appreciative of the creative aspects of his profession, Russ firmly believes that marketing should be regarded more as a science than just art. He regards marketing automation fundamental to ensuring a return on marketing investment.
http://www.linkedin.com/in/rjgould

Paul Gray

As Director of Community Engagement at independent games startup Bubble Gum Interactive, Paul is responsible for marketing and customer experience for Little Space Heroes, a new virtual universe for kids around the world. With over ten years in product marketing, Paul is passionate about focusing product design, development, and delivery around creating outstanding customer experiences.

Sandra Greefkes

Sandra leads client initiatives involving interactive and social media, individualized and permission based marketing, positioning and messaging, brand management, and project management best practices. Prior to joining Filigree, Sandra successfully led best in class business strategies, solutions, process innovation, and integrated campaigns through more than fourteen years of progressive experience in the communications and high-tech industries. Her broad based marketing experience ranges from traditional disciplines to specific expertise using new media approaches to exceed bottom line results.

Bertrand Hazard

Bertrand has over eighteen years of experience devising and executing sales and marketing strategies. In his current role as Sr. Director of Business Strategy, he owns the creation, communication, and execution of the business strategy for SolarWinds, a leading provider of powerful and affordable IT Management software. A seasoned marketing practitioner with a global perspective and entrepreneurial drive, he is a member of the Forrester's Technology Marketing Executive Council, an active board adviser for ProductCamp Austin and a blogger at http://www.arandomjog.com. Twitter: @productmarketer

Christine Heckart

Throughout her twenty-year career, Christine has played an integral role in the marketing and growth strategies of successful technology companies. Before NetApp, she was general manager of marketing for Microsoft's TV, Video, and Music Business. Previously, she was VP and CMO for Juniper Networks, responsible for all aspects of worldwide marketing. An acknowledged industry thought leader, she was named one of Network World's "Top 10 Power Thinkers" and "50 Most Powerful People in the Industry." She has an economics degree from the University of Colorado at Boulder.

Don Jennings

Don is responsible for driving the strategic direction and day-to-day programs such as launches, announcements, product reviews, and analyst and media relations programs in a variety of technology markets. He has helped clients through award-winning, high impact product and service introductions, raising his client's profile in national and majority media outlets. Don has been the lead on more than fifty new service and product introductions for his clients during the past five years.

Reena Kapoor

Reena is Founder and President of Conifer Consulting, a new product and marketing strategy consulting firm. Reena brings over twenty years of new products and marketing experience; including brand management at Procter & Gamble, Kraft Foods, and product management/marketing leadership at Silicon Valley tech start-ups. Reena has a B.Tech degree from Indian Institute of Technology (IIT Delhi), and an MS degree from Northwestern University. Reena completed the 2007 Nike Women's Marathon to become a top Leukemia & Lymphoma Society fundraiser!

Dave Kellogg

Dave Kellogg is a software industry technology executive who specializes in strategy and positioning. Dave is a technology and blogger. Previously, Dave was CEO of unstructured data pioneer MarkLogic, CMO of business intelligence leader Business Objects, and started his career working in technical and marketing roles at Versant and Ingres. You can hear more from Dave on his blog at Kellblog (http://www.kellblog.com) or via his Twitter feed at http://twitter.com/kellblog

Mara Krieps

Mara Krieps is a founder of Pivotal Product Management (http://www.pivotalpm.com), a product management consultancy based in Seattle, WA. She headed the Product Management Consortium (http://www.pmcnw.org) and served on the UW Software Product Management program's advisory board for ten years. She is a recipient of AIPMM's "Excellence in Product Management Education" award, and was named by Puget Sound Business Journal's TechFlash to the "Top 100 Women in Seattle Tech." Mara is an AIPMM Certified Product Manager and Agile Certified Product Manager.

Eric Krock

Eric has nineteen years of experience in product marketing and product management including roles at Netscape, Kontiki, VeriSign, Zvents, and Voximate. *Social Media Marketing Magazine* named him one of the top 85 Chief Marketing Officers on Twitter in April 2011. AgileScout named his blog on Agile product and project management at http://www.voximate.com/blog/ one of the top 200 Agile blogs to follow in 2011. His not-for-profit AIDSvideos.org is the world's leading provider of multilingual HIV/AIDS prevention education videos.

Jeff Lash

Jeff has over ten years of experience in product management and user experience design for online products for companies including Elsevier, MasterCard International, and XPLANE. He blogs about product management at "How To Be A Good Product Manager" (http://www.goodproductmanager.com), runs the product management Q-and-A site "Ask A Good Product Manager" (http://ask.goodproductmanager.com), and dispenses product management wisdom 140 characters at a time on Twitter (twitter.com/jefflash). Jeff lives in St. Louis, MO.

Rich Mironov

Rich Mironov is Principal, Mironov Consulting. He is a serial entrepreneur, product management consultant, veteran of five tech startups, and PM thought leader. His book, *The Art of Product Management*, captures the best of his Product Bytes blog ('01 to '08). He founded the first Product Camp, produced the Agile Alliance's first product manager/product owner conference tracks, and served on the SVPMA board. He often speaks on product organizations, pricing, product strategy, and Agile Transformation.

Dan Olsen

Dan has twenty years of high-tech experience, including five years at Intuit where he led Quicken product marketing. Dan also led product marketing at Friendster. Dan's personalized news startup YourVersion won the TechCrunch50 People's Choice Award. He has a technical background and a Stanford MBA. He has consulted to many startups and is passionate about building and marketing successful, large-scale

products. He regularly speaks at conferences such as the O'Reilly Web 2.0 Expo, P Camp, and Startonomics.
http://www.olsensolutions.com

Joe Pulizzi

Joe is a leading author, speaker and strategist for content marketing. Joe founded the Content Marketing Institute, which includes client-vendor matching site Junta42 and the premier international content marketing event, Content Marketing World, and *Chief Content Officer* magazine. Joe is CEO of SocialTract, the leading service professionals' blogging service and co-author of the highly praised book *Get Content Get Customers*, recognized as the handbook for content marketing, as well as *Managing Content Marketing: The Real-World Guide for Creating Passionate Subscribers to Your Brand.*

Al Scharer

Al leads client initiatives involving thought leadership, customer experience, segmentation, and differentiated marketing. Prior to forming Filigree in 1993, Al held successful positions with high-tech companies for over twenty years. His experience spans the marketing and planning functions of large business, including broad experience in management. Al has provided high-impact facilitation and consultation for executive and middle management at leading high tech companies for close to twenty years.

Dennis Shiao

Dennis is Director of Product Marketing at INXPO and author of the book *Generate Sales Leads With Virtual Events*. At INXPO, Dennis is responsible for go-to-market strategy and execution, and for shaping product and platform evolution via the "voice of the customer." Dennis has managed virtual event campaigns for Cisco, HP, Oracle and Microsoft, among others. Dennis blogs about virtual events at INXPO, and on his personal blog, "It's All Virtual." Dennis can be found on Twitter at *@dshiao*.

Cindy F. Solomon

Cindy is the founder of The Global Product Management Talk (http://www.prodmgmttalk.com). She has over fifteen years' experience in web, services, and software product marketing management and community development at Silicon Valley companies including Apple, Vadem, NetObjects, and start-ups. She holds CPM/CPMM certification.

Adrienne Tan

Adrienne is Director of Brainmates, an Australian product management and marketing agency, she is passionate about helping clients create products that make a difference in the market. At Brainmates she has created successful products as diverse as "not for profit" products to an identity theft protection service. Adrienne has twenty years of product management, business improvement, and operational experience in telecommunications and media, having worked for Australia's largest Internet service provider, Telstra BigPond, and AUSTAR, a Cable operator. http://www.brainmates.com.au | *@brainmates*

Janey Wong

Janey is a Product Marketing professional with nine years' experience in product planning, market strategy development, and customer research. She has worked with international teams, start-ups, and established enterprises in both the corporate and non-profit sectors. Her background is in the new media and entertainment industries working at Brainmates, a product management consultancy, and at Oberon Media, a leader in casual games distribution. She has a M.Com. (Marketing) and a B.Sc. (Environ. Econ. & Policy).

Steven Woods

Steven cofounded Eloqua in 1999 and has held the position of Chief Technology Officer since then. He is responsible for defining Eloqua product strategy and technology vision. Steven's books, *Digital Body Language* and *Revenue Engine*, as well as his blogs and user community support, explore demand generation and current marketing transitions. Prior to cofounding Eloqua, he worked in corporate strategy at Bain & Company and engineering at Celestica. He holds a degree in Engineering Physics from Queen's University.

Natalie Yan-Chatonsky

Natalie Yan-Chatonsky has been building and marketing products and digital applications since the mid-nineties. She has always been on the cusp of innovation, driving major retailers to build e-commerce capability, growing the usage of social networking tools and launching a suite of wireless products. As a brainmates product strategy consultant she has contributed to various stages of the product and marketing planning process from ideation to launch, with leading companies in biomedical, financial, and television sectors over the last three years.

C Product Marketing Resources

The 280 Group website is constantly updated with the latest Product Management and Product Marketing Resources including:

- Free templates, samples and white papers
- Optimal Product Management & Marketing Blog
- Product Management & Product Marketing Books
- Product Management Associations
- Product Management Software Comparison
- Product Management Manifesto
- Product Management & Product Marketing Job Listing Sites
- 280 LinkedIn Product Management & Product Marketing Group (over 23,000 members!)

Visit http://www.280group.com and check the "resources" section for the most up-to-date listings. Also, be sure to subscribe to our free newsletter on our website.

D Product Management and Marketing Templates

The 280 Group also offers Product Management & Product Marketing Toolkits, which include templates, narrated training presentations, and samples. The toolkits can be purchased at http://www.280group.com and cover the following topics:

- Product Roadmap Toolkit™
- Product Launch Toolkit™
- Product Management Lifecycle Toolkit™
- Beta Program Toolkit™
- Product Review Program Toolkit™
- Competitive Analysis Toolkit™
- Developer Program Toolkit™

The 280 Group also makes a number of templates available free for download on the 280 Group website in the "Resources" section under "Free PM Tools," including the following:

- MRD Outline
- Feature Prioritization Matrix
- Beta Program Bug and Feature Database Tools
- AdWords ROI Calculator
- Sample Product Roadmaps
- Developer Program Roadmap
- Developer Program Cost Estimator Tool
- Evangelism Timeline
- Competitive Feature Matrix Comparison Chart
- Product Launch Marketing Budget
- Product Launch Plan Marketing Budget
- Press Release
- Google AdWords Tips and Strategies

280 Group Services

The 280 Group's mission is to help companies deliver products that delight their customers and produce massive profits. We do this by providing consulting, contractors, training, certification, templates, and books in the areas of product management and product marketing. Our Optimal Product Process™ methodology as well as products and services are used by tens of thousands of people worldwide. We also provide comprehensive product management assessments to help companies optimize their process, people and tools.

If you need assistance with a project or need a professional product management or product marketing contractor, contact us for a free no-obligation quote.

Our training and certification programs are available as public courses or self-study courses and can be delivered privately onsite for your company. They include:

- Optimal Product Management & Product Marketing™
- Agile Product Management Excellence™
- Certified Product Manager™ Self-Study Course & Exam
- Agile Certified Product Manager™ Self-Study Course and Exam.
- How to be a Phenomenal Product Manager™
- Certified product Manager Exam Intensive Prep Course
- Social Media Marketing

We also have a wide variety of other courses that can be delivered and range in length from one hour to full days. See our website for additional details.

Appendix

F Endnotes

[i]*Fortune* senior editor Betsy Morris interviewed Steve Jobs as reported in *CNN Money* http://money.cnn.com/galleries/2008/fortune/0803/gallery.jobsqna.fortune/index.html (February, 2008)

[ii]Marketing Sherpa http://www.marketingsherpa.com, "2011 Marketing Sherpa Social Marketing Benchmark Survey." Methodology: Fielded February 2011, N=760

[iii]"Inside Gatorade's Social Media Command Center," http://mashable.com/2010/06/15/gatorade-social-media-mission-control/

[iv]Atul Gawande, "The Checklist," *The New Yorker*, December 10, 2007, http://www.newyorker.com/reporting/2007/12/10/071210fa_fact_gawande#ixzz1XmCgjeso

[v]Danny Sullivan, "Some SEO Advice For Bill Gates," http://searchengineland.com/some-seo-advice-for-bill-gates-34303

[vi]https://adwords.google.com/

[vii]Bruce Clay and Susan Esparza, *Search Engine Optimization All-in-One For Dummies* (Hoboken, NJ: Wiley, 2009), 16–17.

[viii]Hubspot Inc., *HubSpot Essential Step-by-Step Guide to Internet Marketing*, 9.

[ix]Joost de Valk, "Meta Description for Search & Social," http://yoast.com/meta-description-seo-social

[x]"Using Permalinks," WordPress.org Codex,
http://codex.wordpress.org/Using_Permalinks.

[xi]Clay and Esparza, *Search Engine Optimization All-in-One For Dummies*, 346.

[xii]Clay and Esparza, *Search Engine Optimization All-in-One For Dummies*, 342.

[xiii]Joachim Kupke and Maile Ohye, "Specify your canonical," *Google Webmaster Central Blog*, February 12, 2009,
http://googlewebmastercentral.blogspot.com/2009/02
/specify-your-canonical.html

[xiv]Jill Whalen, "The Duplicate Content Penalty Myth," Column: 100% *Organic - Search Engine Optimization Tips*, March 15, 2007,
http://searchengineland.com/
the-duplicate-content-penalty-myth-10741

[xv]Clay and Esparza, *Search Engine Optimization All-in-One For Dummies*, 244.

[xvi]http://www.silverbulletgroup.com/resources/research.shtml

[xvii]Jeff Ernst, Forrester Analyst "Forrester Technology Marketing Council," 2011.

[xviii]Peter F. Drucker, *Management: Tasks, Responsibilities, Practices*, (1974)

[xix]https://adwords.google.com/

[xx]http://www.ContentMarketingInstitute.com

About the Authors

Brian Lawley is the CEO and Founder of the 280 Group. He is the author of four best-selling books, *The Phenomenal Product Manager, Expert Product Management, Optimal Product Process* and *42 Rules of Product Management* and is the former President of the Silicon Valley Product Management Association (SVPMA). He was awarded the Association of International Product Marketing Management award for Thought Leadership in Product Management, and has been featured on World Business Review and the Silicon Valley Business Report. He is the editor of the Optimal Product Management blog and newsletter and also writes guest articles for publications such as the Software Development Forum newsletter, Softletter, and the SVPMA newsletter.

Prior to founding and running the 280 Group, Brian spent many years working on innovative products at world-leading companies, including Digidesign (acquired by Avid), Apple (Product Manager for the MacOS human interface), Claris, Symantec, and Whistle Communications.

Brian is a Certified Product Manager (CPM) and Certified Product Marketing Manager (CPMM) and has a Bachelor's degree in Management Science from the University of California at San Diego with a minor in Music Technology and an MBA with honors from San Jose State University.

Phil Burton has been a consultant and trainer for the 280 Group for four years and has a twenty-five-plus year career in product management and product marketing focused on information security, data communications, and networking. His expertise is in product life cycle management with a focus on customizing the product life cycle process to the specific needs and organizational structure of a company. In addition to executing assignments for specific clients, Phil also conducts most of the 280 Group's Product Management classes. Phil holds a Civil Engineer degree from MIT ("practical PhD"), and is a certified information system security professional (CISSP). Phil is a Certified Product Manager and Certified Product Marketing Manager through the Association of International Product Marketing and Management.

Phil has extensive full life cycle product management and product marketing experience, with a strong understanding of how to tailor the product life cycle process to the specific needs and organizational structure of a client. Phil has expertise in product definition, launch, messaging and positioning, collateral creation, competitive analysis, and sales tool creation. He is also an excellent public speaker for both executive level and technical groups.

With over twenty-five years of marketing experience, **Gary Parker** has a record of creating powerful marketing programs. He has worked with high tech companies to produce innovative marketing programs and collateral that drive demand and support the sales process.

Gary has the strong work skills that are needed to execute smooth, effective cross-functional product launches. He had early, very successful, sales experiences in which he was top sales person every year for four years at Sterling Software and broke company annual sales records twice. This experience helped him migrate into product marketing and provided the foundation for a "hands on" leadership style to closely support sales teams using his product and customer buying criteria expertise.

Gary has an MBA from Golden Gate University, a BS in Business Administration from the University of Arizona and received executive marketing training at Harvard, Columbia and Wharton. He taught Market Planning at the University of Colorado Extension Program.

Gary is a Certified Product Marketing Manager by the Association of International Product Marketing and Management.

He has been a featured conference speaker: Software Business 2009, Laptop Summit 2010, American Institute of Product Marketing and Management 2010 PMEC, P-Camp 2009 Product Marketing session leader and has published articles in Computer Technology Review, CPA Technology Magazine, MIS Week, Software News, Software Magazine plus regular press interviews, blogs and tweets.

Write Your Own Rules

You can write your own 42 Rules book, and we can help you do it—from initial concept, to writing and editing, to publishing and marketing. If you have a great idea for a 42 Rules book, then we want to hear from you.

As you know, the books in the 42 Rules series are practical guidebooks that focus on a single topic. The books are written in an easy-to-read format that condenses the fundamental elements of the topic into 42 Rules. They use realistic examples to make their point and are fun to read.

Two Kinds of 42 Rules Books

42 Rules books are published in two formats: the single-author book and the contributed-author book. The single-author book is a traditional book written by one author. The contributed-author book (like *42 Rules for Working Moms*) is a compilation of Rules, each written by a different contributor, which support the main topic. If you want to be the sole author of a book or one of its contributors, we can help you succeed!

42 Rules Program

A lot of people would like to write a book, but only a few actually do. Finding a publisher, and distributing and marketing the book are challenges that prevent even the most ambitious of authors to ever get started.

At 42 Rules, we help you focus on and be successful in the writing of your book. Our program concentrates on the following tasks so you don't have to.

- **Publishing:** You receive expert advice and guidance from the Executive Editor, copy editors, technical editors, and cover and layout designers to help you create your book.

- **Distribution:** We distribute your book through the major book distribution channels, like Baker and Taylor and Ingram, Amazon.com, Barnes and Noble, Borders Books, etc.

- **Marketing:** 42 Rules has a full-service marketing program that includes a customized Web page for you and your book, email registrations and campaigns, blogs, webcasts, media kits and more.

Accepting Submissions

If you want to be a successful author, we'll provide you the tools to help make it happen. Visit our website at **http://superstarpress.com/** for more information on submitting your 42 Rules book idea.

Super Star Press is now accepting submissions for books in the 42 Rules book series. For more information, email **info@superstarpress.com** or call 408-257-3000.

Other Happy About Books

42 Rules of Marketing

This book is a compilation of ideas, theories, and practical approaches to marketing challenges the author has been collecting over the past 17 years.

Paperback: $19.95
eBook: $11.95

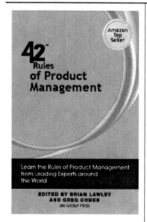

42 Rules of Product Management

With over five hundred years of combined hands-on product management and product marketing experience, the authors each shares one rule that they think is critical to know to succeed in product management.

Paperback: $19.95
eBook: $14.95

42 Rules for Growing Enterprise Revenue

This book is a brainstorming tool meant to provoke discussion and creativity within executive teams who are looking to boost their top line numbers.

Paperback: $19.95
eBook: $14.95

42 Rules to Turn Prospects Into Customers

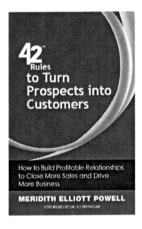

Meridith Elliott Powell draws on her 20-plus years in sales to give you a practical step-by-step guide on how to find the right prospects, build profitable relationships, close more sales and turn your customers into champions for your business.

Paperback: $19.95
eBook: $14.95

Purchase these books at Happy About
http://happyabout.com/
or at other online and physical bookstores.

A Message From Super Star Press™

Thank you for your purchase of this 42 Rules Series book. It is available online at: http://happyabout.info/42rules/productmarketing.php or at other online and physical bookstores. To learn more about contributing to books in the 42 Rules series, check out http://superstarpress.com.

Please contact us for quantity discounts at sales@superstarpress.com.

If you want to be informed by email of upcoming books, please email bookupdate@superstarpress.com.

Lightning Source UK Ltd.
Milton Keynes UK
UKOW041509230312

189485UK00010B/42/P